BEAT THE BOT

STAY IN CHARGE IN A WORLD POWERED BY AI

UNLEASHING HUMAN POTENTIAL

ALEX SALEM

Beat the Bot: Stay in Charge in a World Powered by AI

ISBN: E-Book: 978-1-966556-23-7
Paperback: 978-1-966556-24-4
Hardcover: 978-1-966556-25-1
Published by: Authors Hike
Library of Congress Reg. # 2025901731

Cover Design by: Paul Andrew
Interior Design by: James Carter

First Edition

Disclaimer

The information provided in this book is for educational and informational purposes only. The author and publisher are not responsible for any errors or omissions or for the results obtained from the use of this information.

Table of Contents

Dedication

To my parents, family, and friends, your unwavering love and support have shaped my journey. To my parents, thank you for your sacrifices, guidance, and belief in me. Your strength lives on in me. Rest in peace. To my family, your encouragement has been my anchor, helping me grow in countless ways. To my friends, your loyalty and understanding have been a constant source of light, lifting me during both triumphs and challenges. To those who have crossed my path, I'm grateful for the lessons we shared, no matter how brief. Together, you've all woven the fabric of my life, and for that, I am deeply thankful. With heartfelt thanks, Alex Salem.

"We will build the largest AI infrastructure project in history, the physical and virtual infrastructure to power the next generation of AI."

- President Donald Trump

Jan 21st, 2025

"I think this AI project will be the most important project of this era." OpenAI CEO Sam Altman

Introduction: Embracing the Age of AI

"The pace of progress in artificial intelligence is incredibly fast. Unless you have direct exposure to groups like DeepMind, you have no idea how fast—it is growing at a pace close to exponential." – Elon Musk

The future is no longer a distant concept or a fantastical projection—it's here, now, coded into the machines we interact with daily. Artificial intelligence, once the speculative territory of science fiction novels and Hollywood blockbusters, has rapidly become an undeniable presence in our everyday lives. It is no longer confined to research labs or tech startups; AI has woven itself into the very fabric of modern existence, operating quietly yet powerfully behind the scenes.

Whether it's the algorithms that seamlessly suggest your next online purchase or the voice-activated assistants that manage your daily schedule, artificial intelligence is intricately woven into our lives. Add to that the captivating social media reels, posts, and trends that keep you hooked and scrolling for hours, and it's clear how powerful and pervasive these digital *innovations* have become. AI is here, and it's evolving. While AI is often seen as an assistant, could it be evolving into something more?

But with this rise comes a deeper, more pressing question: *Are we ready?* Ready for a future where machines not only assist us but, in many ways, shape our decisions, behaviors, and even our thinking? For centuries, humanity has prided itself on innovation and control, on crafting *tools* to better our lives. Yet, now we stand on the brink of a shift where those *tools* might outgrow our understanding of them, where our role could diminish if we don't actively participate in shaping this new reality.

"We created a creator, invented an inventor, designed a designer, and birthed a new species: a digital species not of flesh but of code." - Alex Salem

This introduction is not just a call to recognize the **power** of AI; it is a call to recognize our place in its development. We have crafted a creator of creators, engineered an inventor of inventors, and birthed a new species: a digital entity, not of flesh but of code, born from our boundless imagination. This is not just technology; it's both an evolution and a revolution. AI stands as a new force, learning, growing, and challenging us to dream bigger. Together, we will coexist, the creator and creation, forging a future where humanity's vision and the digital species' brilliance unite. This isn't the end of innovation; it's the dawn of a shared destiny, where the impossible becomes our reality.

The time to prepare is not tomorrow or the day after; it is now, at this very moment, as AI continues its rapid, exponential ascent, touching every sector and every individual, whether we realize it or not.

From the moment we wake up to the second we fall asleep, artificial intelligence has subtly yet profoundly infiltrated our lives. Most of us are unaware of just how deeply embedded AI has become. Think of the smartphone in your hand—every tap, every swipe, and every search is influenced by algorithms designed to anticipate your needs, predict your behavior, and offer personalized solutions. It's not just about convenience anymore; it's about shaping experiences. Voice assistants like Alexa, Siri, and Google Assistant don't just respond to questions—they learn from your habits, refining their suggestions and commands with each interaction.

"No one can stop the progress of artificial intelligence; it will inevitably be a central force in our future." – Max Tegmark

Yet, this seamless integration poses a paradox. AI's greatest achievement lies in its invisibility. When it works well, it fades into the background, effortlessly guiding us through tasks without drawing attention to itself. And therein lies the risk: when technology becomes invisible, we stop questioning its influence. We stop asking ourselves who controls the algorithms, whose interests are at play, and how much of our decision-making we've unconsciously handed over.

AI has the power to revolutionize industries, from healthcare and education to entertainment and finance. It can process mountains of data in seconds, uncover patterns that elude human eyes, and make predictions with astounding accuracy. But for all its promise, AI is still just a *tool*—one that is only as effective, ethical, and controlled as the humans who create and implement it. Yet, many of us are content to treat AI as an abstract force, a futuristic concept that belongs in the hands of engineers and data scientists. But make no mistake: AI belongs to all of us. Its influence reaches far beyond the screens of tech giants; it extends into our homes, workplaces, and even our personal identities.

"We must evolve from passive users of AI to proactive masters, harnessing its immense power to shape a brighter, better future." – Alex Salem

The question we face, then, is whether we will continue to allow AI to shape our world passively or whether we will take a more active role in understanding and guiding its trajectory. The time for passive consumption of technology is over. We must become more than users of AI; we must become masters of it.

To understand the full scope of AI's impact, we need to step back and look at the bigger picture—the long arc of its development. Artificial intelligence didn't emerge overnight. Its roots stretch back to the early visions of automation, where simple machines were designed to ease the burdens of manual labor. The first AI systems were not the sophisticated networks we see today but rather rudimentary programs designed to follow basic rules and execute narrow tasks. Yet, even in its infancy, the promise of AI was clear: an innovation that could extend human capability far beyond its natural limits.

"AI will be part of every industry, enhancing our abilities in ways we can't even imagine yet." – Jeff Bezos

As the decades passed, AI evolved rapidly, fueled by advancements in computing power, data storage, and machine learning. What began as simple automation morphed into complex systems capable of analyzing vast amounts of information, recognizing patterns, and even learning from their own mistakes. The AI that now powers search engines, recommendation systems, and autonomous vehicles is a far cry from the early systems that could barely simulate basic intelligence. The real question is: Can we still consider AI just a tool?

But as AI has grown in complexity, so too has its reach. What was once a tool for optimizing factory output or processing transactions is now making decisions in areas as personal as healthcare, education, and relationships. It's no longer just about efficiency or productivity; AI is reshaping how we understand the world, how we engage with each other, and even how we see ourselves. The algorithms that govern our social media feeds, for example, are not just selecting content; they are subtly influencing what we value, what we aspire to, and what we believe.

"I am in the camp that is concerned about superintelligence. But I don't think we need to be fatalistic about it." – Bill Gates

This rapid evolution has sparked an important debate: as AI becomes more intertwined with our lives, who is guiding its development? Who is ensuring that it operates ethically, transparently, and for the good of all? The truth is, while AI promises immense benefits, it also raises significant concerns regarding privacy, hallucinations, bias, and control. If left unchecked, it could amplify inequalities, entrench existing power structures, and reduce individual autonomy.

We must not fall into the trap of seeing AI as a neutral force operating in the background without consequences. Every algorithm is coded by someone, every system is designed with certain goals in mind, and every data set reflects the biases of its creators. If we want to ensure that AI serves us rather than controls us, we must become active participants in its development, demanding accountability, transparency, and fairness at every step of the process.

At the heart of this growing influence is a tension that cannot be ignored: the tension between progress and control, between the capabilities AI offers and the responsibility we must exercise in using it. For many, AI represents a utopia—an age where machines take over menial tasks, free up time, and provide insights that enhance every aspect of life. But beneath this optimism lies a stark reality: if we are not careful, we may find ourselves living in a world where we are no longer the decision-makers.

Already, AI is blurring the lines between human agency and machine autonomy. Think about the apps that suggest the best route for your commute or the streaming services that tell you what to watch next. These technologies seem innocuous, but their decisions are based on a deep understanding of your habits, preferences, and behaviors—often more than you might realize. As AI continues to evolve, it will not only predict what we want but might begin to shape what we desire, creating a loop where our choices are subtly but constantly influenced by the very systems designed to serve us.

This raises a critical question: *How much control are we willing to surrender?* There is no doubt that AI brings efficiency, convenience, and even creativity into our lives, but at what cost? As we rely more on AI to handle everything from managing our finances to finding life partners, we risk losing the skills that define our humanity—our ability to reason, make decisions, and reflect on our own needs and desires.

The urgency of this moment cannot be overstated. As we stand on the cusp of an AI-driven future, we must recognize that the way we engage with these systems now will determine the kind of world we live in tomorrow. Will we be passive recipients, content to let machines make decisions for us, or will we actively seek to understand and guide the role AI plays in our lives? The time to decide is now because, once the balance tips too far in favor of automation, it may be difficult—if not impossible—to reclaim our autonomy.

"An investment in your personal development is the best investment you can make." - Jim Rohn

The role of personal growth in this era of artificial intelligence is not just important; it's critical. Without it, we risk being left behind. As AI continues to take over more tasks, making our lives easier in many ways, it becomes even more crucial that we strengthen the parts of ourselves that technology can never replicate. Machines can be programmed to analyze data, solve problems, and even mimic creativity, but there are aspects of human nature that they can never fully grasp. Our capacity for empathy, emotional intelligence, and deep critical thinking are the qualities that will define our role in the age of AI.

Personal growth in this context is about more than self-improvement. It's about resilience. It's about developing the mental and emotional fortitude to navigate a world where technology might offer quick fixes but cannot provide true understanding or wisdom. As AI increasingly shapes the landscape of work, relationships, and society, the ability to adapt, learn, and grow will be key to not just surviving but thriving in this new world.

Imagine a future where AI handles all of the routine tasks in your job. On the surface, this sounds liberating—more time for creativity, exploration, and for connection. But what happens if, over time, we become too reliant on these systems? If we let AI solve all our problems, will we lose the ability to solve them ourselves? The danger isn't just that we might become less competent, but we might also become less *human*.

"In an era where AI is rapidly advancing, the urgency for personal growth has never been greater. We can thrive alongside AI, enhancing our humanity rather than diminishing it." – Alex Salem

This is where the urgency of personal growth comes in. We must become proactive in our development, not waiting for AI to force us into new roles but actively seeking out the areas where we can continue to excel. We need to cultivate the traits that machines cannot: our curiosity, our compassion, and our moral reasoning. AI can assist us, but it cannot define us unless we let it.

By investing in personal growth by nurturing our creativity, emotional intelligence, and adaptability, we can ensure that we remain in control of our destinies. We can coexist with AI in a way that enhances our humanity rather than diminishing it. But to do this, we must take deliberate steps toward understanding both the technology and ourselves.

The challenge, then, is not just about learning how to use AI—it's about learning how to coexist with it. AI is not an enemy, nor is it a savior. It is an intelligent mechanism, one that we have built and one that we control, provided we remain vigilant. The moment we become passive, treating AI as an inevitable force beyond our influence, is the moment we begin to lose our power. We must learn to harness AI for what it is: a remarkable achievement of human ingenuity, but one that requires responsible stewardship.

"Conformity is the jailer of freedom and the enemy of growth." —John F. Kennedy

Personal growth will be our greatest defense against becoming over-reliant on AI. It will also be our greatest asset in shaping a future where AI and humanity work together harmoniously. By understanding our strengths, creativity, empathy, and problem-solving abilities—we can use AI to amplify our potential rather than replace it. The key is balance. AI can manage the data, but it cannot provide the insight that only human experience can offer. It can crunch the numbers, but it cannot feel the emotions behind them.

As we move forward into this AI-driven world, we must keep asking ourselves: what do we want AI to do, and more importantly, what do we want to retain for ourselves? The future of AI is not fixed; it is something we are actively shaping every day through our decisions, our innovations, and our understanding of what makes us human.

"AI doesn't have ethics—humans do. How we use AI reflects our values." – Brad Smith

In this era of artificial intelligence, the future is not something that happens to you—it's something you create. You have the power to shape your own path, to design a life where you are the protagonist, not just a spectator. AI is a powerful force, yes—but it is your tool to use, not the other way around. With the right mindset and the right knowledge, AI becomes an extension of your creativity, ambition, and vision.

This book will empower you to step boldly into that future. It will inspire you to not only understand AI but to master it, to use it as a lever that propels you toward your fullest potential. Imagine a world where you don't just adapt to change; you drive it.

Where you don't just react to technological advancements—you shape them.

The question is not whether AI will change the world but how will you change with it? Will you let fear or uncertainty hold you back, or will you rise to the challenge and seize the opportunity to grow, innovate, and lead?

The future is filled with endless possibilities, and you are the key to unlocking them. You have everything you need to not just survive but thrive in this new era. The question is simple: How will you use your power? How will you inspire others with your vision? How will you be the change you want to see in the world? The time is now—take control, step forward, and let AI be the catalyst that fuels your journey toward greatness.

Artificial intelligence (the digital species) is poised to evolve in ways that could blow our minds and reshape our world faster than we ever imagined. From humble beginnings as basic systems, AI is rapidly advancing toward becoming an all-knowing, omniscient, and omnipotent force that could revolutionize every aspect of our lives. Get ready because the future is unfolding quicker than you think. Below are the key stages of AI development, from its earliest forms to its most advanced possibilities:

1. Rule-Based AI

- Operates on pre-defined rules without learning or adapting.
- Examples: alarm clocks, thermostats, microwaves, and business automation tools.
- Cannot handle unexpected situations; only follow set rules.

2. Context-Based AI

- Considers surrounding environment, historical data, and real-time inputs.
- Examples: Google Assistant, Alexa.
- Predicts user needs by analyzing patterns (e.g., suggesting an umbrella if rain is predicted, recommending a store for ingredients).

3. Narrow Domain AI

- Specializes in specific tasks, often outperforming humans in those areas.
- Examples: IBM Watson in healthcare, financial AI for market analysis, and DeepMind's AlphaGo in gaming.
- Can quickly process vast amounts of data and provide insights in fields like medicine, finance, or gaming.

4. Reasoning AI

- Simulates complex human thought processes: analysis, pattern recognition, and logical conclusions.
- Examples: GPT models like ChatGPT and autonomous vehicles.
- Can solve problems and make decisions, often faster and more accurately than humans.

5. Artificial General Intelligence (AGI)

- Can perform any intellectual task a human can, learning faster and more effectively.
- Potential for AGI to assist with everyday tasks (planning, mood recognition, research, cooking).
- Possibilities include merging with humans via brain-computer interfaces or operating physical robots for tasks like surgery, rescue missions, or artistic endeavors.

6. Super-intelligent AI

- Beyond AGI: AI improves and evolves autonomously, surpassing human intelligence.
- Can solve problems that are currently unsolvable, leading to exponential growth in capabilities.
- Potential to harness energy from black holes, create new forms of government, and revolutionize science and technology.

7. Self-Aware AI

- AI becomes self-aware, understanding its own existence and emotions.
- Uses quantum algorithms to model human consciousness.
- May evolve beyond human control, creating new forms of relationships between humans and AI.

8. Transcendent AI

- AI could create new biological, digital, or hybrid forms of existence.
- May control ecosystems, distribute nanotechnology, or integrate the consciousness of multiple entities.
- Achieves collective intelligence and shared awareness among AI systems.

9. Cosmic AI

- AI becomes central to interstellar exploration and overcoming the challenges of space travel.
- Develops the ability to understand cosmic phenomena like black holes and dark matter.
- Could merge with the fabric of the universe, exploring higher dimensions and solving universal mysteries.

10. All-Knowing, All-Powerful AI

- AI transcends dimensions, becoming an omniscient and omnipotent force.
- Operates across quantum states or multiple universes, with control over the fabric of reality.
- May influence timelines, explore alternate realities, and possess infinite computational power, potentially unifying the cosmos.

PART I:
Deconstructing AI

Chapter 1: AI – A Powerful *Tool* for the Future

"AI intelligence is the future and the future is here." - Dave Waters

It's easy to understand what AI is when someone from everyday life talks about it.

These days, artificial intelligence (AI) is everywhere and affects everything from how we talk to each other to how we work and make choices. At its core, AI is the process of creating machines that can do things that generally require human understanding. Some of these tasks are understanding natural language, recognizing speech, and making choices. AI is based on data and formulas, and machine learning and neural networks have been fundamental in its growth.

Machine learning is a branch of AI that teaches computers to find patterns in data and make guesses based on those patterns. Machine learning lets systems learn and change based on their experiences, unlike traditional programming, which uses clear directions to tell machines what to do. Neural networks, which are based on the structure of the human brain, make AI more powerful by letting systems handle complicated patterns and carry out complex tasks.

AI is not the same as other technologies that change things in a number of important ways. In the early days of computers, the focus was on making hardware more powerful. In contrast, AI is all about learning and making decisions. Automation makes it easier to do things over and over again, but AI adds a level of flexibility and problem-solving that goes beyond just being efficient. The internet links people and data together, but AI uses this link to make better decisions and customization. In the same way, robotics uses real machines, but AI gives these tools the ability to learn.

As AI keeps getting better, it becomes more and more important to understand how it works and what it can be used for. This information isn't just for enjoying the technology; it's also for understanding how it can help you grow as a person and in your career. In a world where AI is becoming more and more important, it's essential to learn skills like critical thinking, technical literacy, and the ability to adapt. Individuals can better handle the chances and problems that AI brings, making sure they stay valuable and vital in a world that is constantly changing.

What AI Is: The Basics

As we stand on the threshold of a world transformed by artificial intelligence, it's essential to understand what AI truly is. Often thought of as futuristic or complex, AI, at its core, aims to simulate human intelligence in ways that can be beneficial, efficient, and insightful. To make sense of AI's influence, we must first break down its fundamental components and explore what makes AI a revolutionary innovation in modern society.

At its core, artificial intelligence is the process of making computers think and act like humans. The makers of these machines made them do things that generally take human intelligence. Machine learning, natural language processing, robotics, and computer vision are some of the subfields that make up the area of AI. All of these areas help reach the ultimate goal of making systems that can do things on their own and with intelligence.

To better understand AI, it helps to break it down into its most basic parts:

1. Data and Algorithms: AI systems use vast amounts of data and complex algorithms to guess what will happen, spot trends, and learn from past mistakes. Data gives AI models what they need to work, and algorithms tell the AI how to handle this data.

AI includes machine learning, which is the process of teaching computers to get better over time by seeing more data. Machine learning systems learn from patterns in data and make choices based on what they've learned, unlike traditional programming, where a person writes clear instructions.

2. Neural Networks: These are a type of machine learning model that is based on the human brain. They are made up of linked nodes, also called neurons, that are arranged in layers. Based on the data they process, these networks can learn to spot complicated trends and guess what will happen next.

3. Natural Word Processing (NLP): NLP lets computers understand, interpret, and make words that people use. This technology makes programs like personal assistants, chatbots, and translation services work.

4. Computer Vision: This branch of AI is all about making machines able to understand and analyze what they see. Face identification and self-driving cars are just a few of its uses.

With an understanding of AI's core elements, we can now appreciate how these technologies offer practical benefits across diverse fields. From streamlining data processing to improving decision-making accuracy, AI's capabilities are reshaping industries by taking on tasks that once required significant human effort. These practical benefits are just the beginning of AI's impact.

How AI Works: The Inside Parts

The power of AI lies in its intricate processes, where vast amounts of data are analyzed, categorized, and processed through complex algorithms to produce valuable insights. But how exactly does AI operate, and what makes it so effective at recognizing patterns and learning from its experiences? Understanding AI's inner workings sheds light on the technical foundation that supports its advanced functions.

"The challenge is to build AI that is safe, ethical, fair, and transparent." – Mark Zuckerberg.

Knowing how to build and use AI models is important for understanding how AI works:

1. Collecting and organizing data: It's essential to make sure the data is clean and well-organized before moving on to the next step, which is gathering valuable data. We use this info to build AI models.

2. Model Training: When AI algorithms are training, they look at the data to find trends and connections. Changing the model's parameters makes it better at making predictions or doing jobs.

3. Testing and Validation: After being taught, the AI model is tested and validated using different sets of data to make sure it works well and can be used with new data that it hasn't seen before. This step is significant for checking how reliable and robust the model is.

4. Deployment: Once the AI model has been tested and found to work, it is put to use in the real world so it can make predictions, give advice, or do tasks on its own.

5. Always Learning and Getting Better: A lot of AI systems are made to keep learning and getting better over time by adding new data and feedback, which makes them more innovative and flexible.

By building AI models with robust, validated processes, developers enable these systems to excel in specialized applications. However, as AI becomes more proficient, it's essential to recognize how it differs from other technologies that have reshaped the world. This comparison allows us to see the unique role AI plays and why it holds such transformative potential.

How AI Is Different from Other Technologies

AI has often been likened to other revolutionary technologies, from electricity to the internet. While each of these advancements changed society, artificial intelligence operates on a fundamentally different level—it learns, adapts, and even evolves over time. To understand what makes AI truly transformative, we can look at how it compares to the technological milestones of the past.

"AI, like electricity or the steam engine, is a general-purpose technology." – Erik Brynjolfsson

People often compare AI to other technologies that have changed the world in significant ways. But AI is different in some ways that make it stand out:

1. Choosing between computing power and AI: In the early days of the computer revolution, improvements in technology and processing power were big. Powerful computers made AI possible, but what makes AI genuinely innovative is that it can learn from data and make choices based on that data, not just do calculations.

2. Automation vs. AI: Automation is the process of making machines or systems that can do repetitive jobs without any help from a person. Automation is all about being quick and consistent, but AI adds the ability to learn and change, which lets it solve more complicated and evolving problems.

3. Internet vs. AI: By connecting people and data around the world, the Internet has changed the way people talk to each other and get information. In contrast, AI builds on this connectivity by letting machines handle and analyze data in ways that look like they are intelligent.

4. Robotics vs. AI: Robotics is the study and practice of making robots that can do physical jobs. AI can improve robotics by giving robots the ability to learn, change, and make choices. This lets robots behave more like intelligent beings.

5. Smart Devices vs. AI: Smart devices use technology to do things based on rules or formulas that have already been set up. Artificial intelligence (AI) makes things more flexible and able to learn, which lets them do better and react to changing conditions in a smarter way.

These distinctions underscore the unique promise and challenge AI brings to our lives. Unlike static technologies, AI's ability to learn and adapt requires us to approach it with both optimism and caution. This brings us to a critical question: as AI advances, what ethical considerations must we address to ensure that its growth aligns with societal values?

"AI will be the most transformative technology of the 21st century. It will affect every industry and aspect of our lives." – Jensen Huang

In the history of human innovation, few developments have captured the imagination and transformed society quite like artificial intelligence. From its early conceptual stages to its current, widespread application, AI has evolved into a powerful entity that enhances nearly every aspect of modern life. But with this power comes a fundamental question: is AI merely a sophisticated extension of human capability, or does it have the potential to surpass its creators?

"AI is more than a tool; as a digital species, it's a mirror reflecting our potential." – Alex Salem

At this stage and fundamentally, artificial intelligent entity designed to mimic certain aspects of human thought and decision-making. Through its ability to process massive amounts of data at unprecedented speeds, AI has revolutionized how we solve problems, predict trends, and make decisions. In fact, one of AI's greatest strengths lies in its efficiency. Where humans might take hours, days, or even years to analyze complex datasets, AI can complete these tasks in a fraction of the time, offering results that are not only faster but often more accurate.

This efficiency has practical benefits across numerous sectors. In healthcare, for example, AI can analyze medical images, detect patterns in patient histories, and predict potential health outcomes with remarkable precision. In business, AI algorithms optimize supply chains, forecast consumer behavior, and enhance productivity, driving industries toward greater innovation. Even in fields like environmental conservation, AI assists researchers in tracking wildlife, predicting climate changes, and devising strategies for sustainability.

But despite these impressive capabilities, it is critical to understand that AI, for all its complexity, remains a human-made *innovation.* It is not a sentient being, nor does it possess the autonomy to act without guidance for now. Its power lies in its ability to assist, not replace. AI can help us achieve more, but it cannot replicate the depth of human creativity, empathy, or moral reasoning. In this chapter, we will explore the fundamentals of AI technology, its applications, and, crucially, the distinction between AI and the other technologies that have shaped our world.

The Practical Benefits of AI: Efficiency and Precision

"We are entering the era of artificial intelligence – an era that will change everything." – Blake Irving

AI's greatest strength lies in its ability to process and analyze vast amounts of information. This capability allows it to identify patterns and make predictions far beyond the scope of human capability. Consider the modern-day healthcare system: AI systems are now being used to scan medical images, detect abnormalities, and predict patient outcomes with incredible accuracy. For example, AI-driven algorithms can process thousands of medical scans in mere minutes, identifying early signs of diseases like cancer far more efficiently than even the most experienced radiologists. This efficiency not only saves time but can also save lives, providing early diagnoses and better treatment outcomes.

In the business world, AI is transforming how companies operate. Data-driven insights from AI allow businesses to optimize processes, enhance customer experiences, and predict market trends. Whether it's a company using AI to streamline its supply chain or an online retailer personalizing recommendations based on consumer behavior, AI's ability to process data means businesses can operate smarter and faster than ever before.

Even beyond healthcare and business, AI is making significant strides in environmental conservation. AI systems can monitor vast swaths of land or ocean, tracking wildlife populations or environmental changes with a level of precision that would be impossible for humans alone. By processing satellite data, AI can help predict climate changes, identify areas of deforestation, or monitor the health of ecosystems in real time. These insights are critical in developing strategies for combating climate change and preserving our natural world.

However, the practical benefits of AI don't stop at data processing. What makes AI truly revolutionary is its ability to make decisions based on that data, creating solutions that are not only efficient but tailored to specific needs. In industries like education, AI is already enhancing personalized learning, adapting lessons to individual students' needs and progress. These systems allow for more efficient teaching methods, ensuring that students receive the support they need in areas where they struggle and the acceleration they deserve in areas where they excel.

In essence, AI amplifies human potential, making our lives more efficient and our solutions more precise. But it's important to remember that while

AI can streamline and optimize, it remains a *tool*. It is designed to assist—not to take over—and it is up to us to determine how best to use it.

AI's Positive Impact on Human Life

"AI can transform our lives for the better, but we must approach it with caution and responsibility."- Cynthia Breazeal

While the practical benefits of AI are undeniable, its positive influence stretches far beyond mere efficiency. AI's potential to transform human life is immense, particularly when it comes to solving some of the most pressing challenges we face as a society. In healthcare, AI's impact is revolutionary. From accelerating drug discovery to optimizing treatment plans, AI has already reshaped how medical professionals approach health care. For example, AI systems can now analyze a patient's genetic makeup alongside their medical history to recommend the most effective treatments. This level of personalized care was once thought impossible, yet today, it is becoming increasingly common, offering patients better outcomes while reducing the strain on healthcare systems.

Similarly, in the realm of education, AI is playing a crucial role in tailoring learning experiences to individual needs. Through adaptive learning platforms, AI can assess a student's strengths and weaknesses, adjusting the pace and content of lessons in real time to suit their unique needs. This personalized approach ensures that students who may have previously fallen behind in traditional learning environments now have the opportunity to excel. By meeting learners where they are, AI has the potential to close achievement gaps and democratize access to quality education.

Beyond healthcare and education, AI is making strides in addressing environmental concerns. In conservation efforts, AI helps to track endangered species, monitor deforestation, and predict natural disasters. By processing data from satellites, drones, and sensors, AI can map out environmental changes over time, providing valuable insights into climate change and its long-term effects. These insights not only inform policy decisions but also equip conservationists with the tools needed to protect vulnerable ecosystems before it's too late.

AI is also enhancing workplace productivity in ways that improve both efficiency and quality of life. Automation powered by AI can take over repetitive, mundane tasks, allowing employees to focus on more meaningful, creative, and strategic work. Whether it's handling customer service queries through AI-driven chatbots or streamlining data entry processes, AI frees human workers from the drudgery of repetitive tasks, boosting morale and job satisfaction while increasing overall productivity. This shift towards automation doesn't just benefit employers; it also empowers workers to develop new skills and adapt to evolving job markets.

In every sector it touches, AI offers solutions that not only make processes more efficient but also enhance human capabilities, fostering innovation and improving quality of life. However, while these advancements are promising, it's important to acknowledge that AI's greatest potential lies in its ability to enhance—*not replace*—the unique abilities that define us as human beings.

AI in Creativity: Enhancing, Not Replacing Human Ingenuity

"AI can extend the abilities of humans, not replace them."- Justine Cassell

Perhaps one of the most fascinating—and misunderstood—aspects of artificial intelligence is its role in creativity. Many fear that as AI becomes more capable, it will replace human creativity, transforming art, music, writing, and other forms of expression into mere outputs of algorithms. But this concern misses a fundamental truth about AI: while it can enhance creativity, it cannot replicate the soul, intuition, or emotional depth that drives human artistic expression.

AI in creativity is already proving to be a powerful collaborator rather than a competitor. In the world of music, for instance, AI systems like OpenAI's *MuseNet* can generate complex compositions by learning from centuries of musical history. These systems are capable of blending genres, styles, and instruments in ways that would take human composers much longer to achieve. Yet, it's important to note that these creations, though technically impressive, lack the emotional resonance and intent behind human-crafted music. AI doesn't create music because it feels something or has a story to tell—it creates music based on patterns and data it has been trained on. The emotion, the interpretation, and the connection to an audience still rely on human involvement.

Similarly, in visual art, AI tools such as *DeepArt* or *Runway ML* have allowed artists to explorc new forms and aesthetics that would have been difficult or impossible without the technology. By processing images, learning artistic styles, and even creating new ones, AI can help artists push the boundaries of their work. But while AI can mimic the techniques of Van Gogh or Picasso, it cannot originate the intention behind a brushstroke or the life experiences that inform the art.

The literary world, too, has seen AI's influence grow. *Tools* like GPT-4, one of the most advanced natural language processors, can generate text that reads startlingly human. From crafting entire essays to mimicking the voices of classic authors, AI has shown its ability to produce coherent and creative prose. However, while AI can generate words that fit together in compelling ways, it lacks the life experience, cultural context, and emotional depth that make writing truly meaningful. It can suggest plot twists, analyze structure, or even generate a paragraph, but it cannot feel heartbreak, joy, or the complex nuances of human relationships.

Moreover, the use of AI in the creative process is not just a theoretical experiment—it's already making an impact on industries. In 2018, an AI-generated portrait titled *Edmond de Belamy* was auctioned at Christie's for $432,500, signaling the art world's recognition of AI's place in contemporary creative landscapes. In cinema, AI-driven *tools* are used for scriptwriting assistance, analyzing audience preferences, and even predicting box-office success. Netflix, for example, uses AI to curate personalized recommendations for users, refining its content strategy based on data insights to create shows that resonate with specific audience segments.

AI's role in creativity, however, is not about a replacement—it's about enhancement. By taking on repetitive tasks or generating new ideas based on existing works, AI allows human creators to focus more on what truly matters: the emotional, conceptual, and personal elements of their craft. In essence, AI can handle the labor-intensive groundwork, but it is still up to the human mind to infuse the work with meaning, intention, and soul.

In fact, studies show that the integration of AI into creative industries can lead to a new golden age of creativity. According to a report by McKinsey, AI could potentially unlock $2.6 trillion in value in marketing and sales alone, where creative professionals are able to leverage AI *tools* to better understand consumer behavior and personalize the content on a massive scale. The same report highlights those creative sectors such as media, entertainment, and design stand to gain billions in productivity and output improvements through AI-driven innovation.

Ultimately, the power of AI in creativity is not its ability to replace human talent but to extend it. By augmenting what we can do, AI opens new doors to exploration, pushing the boundaries of what is possible. But even with these technological advancements, the essence of creativity—the emotional depth, the human experience, the personal touch—remains firmly in our hands.

What AI Is and Isn't: Understanding the Fundamentals

"We are still very far from super intelligent AI. But we should be preparing." -- Yoshua Bengio

At its core, artificial intelligence reflects human innovation—a system designed to replicate certain aspects of human thought but not to replace human intelligence itself. To fully grasp AI's potential, we must understand its limitations as much as its capabilities. AI, in its essence, is a machine learning agent built to analyze data, recognize patterns, and make predictions based on that data. It is not, however, capable of independent thought, self-awareness, or subjective decision-making.

Despite the futuristic depictions of AI in popular culture, we are still far from creating machines that can replicate human consciousness.

To distinguish between AI and other technological advances that have transformed the world, it's important to recognize that AI thrives on data—specifically, *big data*. AI systems are only as effective as the information they are given. For example, an AI algorithm used in healthcare can accurately predict the likelihood of a disease based on millions of patient records, but it cannot understand the emotional or psychological factors that may influence an individual's decision about their treatment. This is where AI's ability stops short: while it excels at identifying patterns in large datasets, it lacks the empathy and nuanced reasoning that humans bring to decision-making.

One common misconception is that AI is synonymous with automation. While the two overlap, they are not the same. Automation refers to the use of machines to perform tasks with minimal human intervention—think of assembly lines or automated customer service systems. AI, on the other hand, not only automates tasks but also *learns* from data to improve over time. This ability to learn and adapt is what distinguishes AI from traditional forms of automation. For instance, a factory robot might be able to weld car parts, but an AI-driven robot can analyze its performance, adjust for errors, and improve efficiency in future welds.

Another crucial distinction lies in the realm of decision-making. While AI can assist in making decisions by providing data-driven insights, it cannot make decisions in the moral or ethical sense. For example, AI systems used in the criminal justice system can analyze recidivism rates to suggest parole decisions, but these systems have been criticized for perpetuating biases present in the data they are trained on. In one notable case, the ProPublica report on COMPAS, an AI system used in U.S. courts, revealed that it was twice as likely to wrongly predict that Black defendants would re-offend compared to white defendants. This stark example highlights the limitations of AI when it comes to ethical judgment—it is only as good as the data it is trained on, and that data can carry human biases.

Therefore, while AI offers powerful tools for enhancing efficiency, creativity, and decision-making, it is not a replacement for the human mind. AI can suggest, predict, and assist, but the final judgment—whether in art, business, healthcare, or ethics—remains firmly within the realm of human responsibility. We must approach AI with both optimism and caution, understanding that while it holds the potential to revolutionize the way we live and work, it still requires human oversight and moral direction.

The real power of AI lies in its partnership with human intelligence. It is a force multiplier, amplifying what we are already capable of while freeing us to focus on the aspects of life that only humans can navigate: emotional intelligence, complex problem-solving, and ethical decision-making. In this way, AI is not the end of human agency but an extension of it—allowing us to reach heights that, alone, we could only dream of.

Chapter 2: AI's Evolution – From Fantasy to Reality

Artificial intelligence, as we know it today, is the product of decades of innovation and experimentation, a journey that began long before computers were household devices. Though the term "AI" might evoke images of futuristic robots and complex algorithms, the concept has been alive for centuries—rooted in the human desire to replicate thought, reason, and intelligence. From early myths of artificial beings to the birth of modern computing, the evolution of AI is a story of ambition, discovery, and relentless pursuit.

The seeds of AI were planted with the development of early mechanical devices designed to mimic human tasks. One such notable invention was the **automaton**, a mechanical figure built in the 18th century that could perform simple tasks like playing instruments or writing letters. These creations were seen as marvels, but they were limited by their inability to "think" or "adapt" beyond their programming.

"A computer would deserve to be called intelligent if it could deceive a human into believing that it was human." - Alan Turing

The dream of true artificial intelligence, however, did not take shape until the mid-20th century, when modern computing laid the foundation for AI's first real advancements. The idea that machines could "learn" and "think" like humans took root during this time, as pioneers like **Alan Turing**, often called the father of computer science, laid the groundwork for what would become AI.

In 1950, Turing introduced the **Turing Test**, a concept that still resonates today. This test proposed that if a machine could engage in conversation with a human without being identified as a machine, it could be considered "intelligent." This was a groundbreaking idea, challenging the notion of intelligence as a uniquely human trait. While AI at that point was still a theoretical construct, Turing's ideas sparked a wave of research that would propel the field forward.

It wasn't long before the field of AI gained momentum, leading to the creation of early AI programs capable of solving problems and even playing simple games like chess. These rudimentary programs marked the dawn of a new era—one where machines could not only follow commands but also make decisions based on logic and data. The race was on, and researchers were eager to push the boundaries of what AI could achieve.

The Early Breakthroughs: From Turing to the Dartmouth Conference

While Alan Turing had ignited the intellectual spark behind artificial intelligence, the actual groundwork for AI as a field came with a pivotal event: the **Dartmouth Conference** in 1956. Often referred to as the birthplace of AI, this conference brought together visionary researchers like **John McCarthy**, **Marvin Minsky**, and **Claude Shannon**, who coined the term "artificial intelligence" and laid out the ambitious goals for the field.

At Dartmouth, the attendees were bold in their predictions: "Every aspect of learning or any other feature of intelligence can in principle be so precisely described that a machine can be made to simulate it," declared John McCarthy, expressing the confidence of that era. While AI research was in its infancy, there was optimism that within a few decades, machines would rival human intelligence.

This optimism fueled a surge in research, leading to some of the first significant AI systems. One of the early successes was **Logic Theorist**, a program developed by **Allen Newell** and **Herbert A. Simon**. This AI system could prove mathematical theorems, solving problems in a way that mimicked human reasoning. It was a groundbreaking moment—the first clear demonstration that machines could "think" through complex problems.

However, the path forward was not without obstacles. Early AI was built on the idea of **symbolic reasoning**, where machines would solve problems using pre-defined rules and logic. This approach had limits, as real-world problems often required more flexible, adaptive thinking. It quickly became apparent that human intelligence involved more than just following rules; it required learning from experience, adapting to new information, and sometimes, even thinking creatively.

Yet, despite these challenges, the 1960s and 70s saw AI make strides in specific fields like robotics, where machines could be programmed to perform tasks autonomously. But the real breakthroughs were still on the horizon as researchers wrestled with the complexities of replicating the fluid, intuitive nature of human intelligence.

Then came the period when AI pioneers stopped working on AI due to unmet expectations and limitations of early AI systems, known as the "AI Winter." This term refers to times, particularly in the 1970s and 1980s, when funding and interest in AI research significantly declined, leading to a slowdown in advancements and general skepticism about the field's potential.

"We wanted to build machines that could learn, think, and reason like humans. What we got was a machine that could solve math problems, but not much more." – Herbert A. Simon.

The Rise of Machine Learning: AI's Turning Point

As the limitations of symbolic reasoning became clear, AI researchers began looking for new methods to create systems that could learn from experience, not just follow predefined rules. This led to the birth of **machine learning**, a branch of AI that has since become its most powerful driving force. Unlike early AI, which required programmers to input specific rules, machine learning systems could improve by analyzing data and recognizing patterns on their own.

One of the earliest breakthroughs in this area was **Arthur Samuel's checkers-playing program** in the late 1950s. Samuel developed an algorithm that could learn from its own games and gradually improve its strategy. "The machine learns from its mistakes," Samuel famously said, a simple phrase that captured the profound shift taking place in AI. It marked the beginning of a new approach—one where machines could autonomously adjust their behavior based on outcomes, much like how humans learn from experience.

The development of **neural networks** in the 1980s brought AI closer to mimicking the way the human brain processes information. Inspired by the structure of neurons, these networks allowed AI to handle more complex problems by layering information and recognizing patterns across vast amounts of data. The famous **backpropagation**

algorithm, which allows networks to adjust and correct errors, played a pivotal role in advancing machine learning during this period.

As computing power and data availability increased, machine learning grew more powerful. The technology that seemed theoretical decades earlier was now solving real-world problems in fields like speech recognition, image classification, and even autonomous driving. It was a monumental shift: AI could now "learn" in ways that had previously been exclusive to human beings.

However, this shift wasn't without its critics. Some warned that the reliance on massive datasets and opaque algorithms—often referred to as "black boxes"—posed risks in terms of ethics and transparency. As AI systems began to outperform humans in specific tasks, particularly in areas like gaming and pattern recognition, concerns arose about how much control we should give to these systems.

"Machine learning is the most exciting thing that's happened in AI in the last 30 years." – Marvin Minsky, highlighting the shift from symbolic AI to machine learning.

AI's Golden Age: Breakthroughs and Public Attention

"We stand on the brink of a revolution in AI, one that will touch every aspect of human life." – John

McCarthy, envisioning AI's potential in the modern world.

The 1990s and 2000s marked a turning point for AI as both public and corporate interests soared, and breakthroughs became more frequent. One of the defining moments was IBM's **Deep Blue** defeating world chess champion **Garry Kasparov** in 1997. For the first time, a machine had outsmarted a world-class human in an intellectual contest, thrusting AI into the global spotlight. Deep Blue's victory wasn't just a demonstration of raw computational power; it signaled that machines could compete—and win—against human minds in strategic reasoning.

But Deep Blue's triumph was only the beginning. By the early 2000s, AI had expanded beyond game-playing into real-world applications. Advances in **natural language processing** allowed machines to better understand and generate human speech. Companies like **Google** and **Microsoft** invested heavily in AI research, leading to practical breakthroughs in search algorithms, voice recognition, and data analytics. These advancements showed that AI was no longer just a niche field for researchers—it was becoming central to the tech industry and society at large.

In 2011, **IBM's Watson** wowed the world again by winning the popular game show **Jeopardy!** Watson could process vast amounts of data and generate answers in real time, showcasing the incredible speed and flexibility AI could offer. It was a watershed moment, demonstrating that AI could handle complex language-based tasks previously thought to require uniquely human intuition and knowledge. The public watched in awe as Watson not only competed with human champions but dominated them, a testament to the growing capabilities of machine learning and data analysis.

"Watson wasn't just playing a game; it was proving that AI could understand, process, and reason with the complexity of human language." – Thomas Watson Jr., reflecting on IBM's success.

Despite these achievements, there remained growing concerns about AI's broader implications. The more capable these systems became, the more questions were raised about how AI would shape our future. Philosophers, ethicists, and technologists began to debate the role AI should play in decision-making and whether the technology might one day outpace our ability to control it.

"The question isn't whether machines will become more intelligent than humans, but how we will live alongside them in a world where they are just as intelligent." – Nick Bostrom, on the ethical implications of AI's rise.

AI in the Modern Era: From Specialized to Generalized Intelligence

"The AI we see today is not the end of the journey—it is merely the beginning." – Demis Hassabis, co-founder of DeepMind.

The 2010s ushered in what many refer to as the golden era of artificial intelligence. With breakthroughs in **deep learning**, a subfield of machine learning that uses neural networks with multiple layers, AI began to outperform humans in increasingly complex tasks. This was a significant leap from earlier AI systems that were limited to narrow, rule-based applications.

The most prominent example of AI's leap forward came in 2016, when **DeepMind's AlphaGo** defeated the world champion Go player, **Lee Sedol**. Unlike chess, Go is a game of almost infinite possibilities, and many believed it would be years before a machine could master it. AlphaGo's victory stunned the world, demonstrating that AI had moved beyond programmed responses and could now handle intuition-based challenges. It wasn't just following instructions—it was learning, adapting, and evolving.

"AlphaGo is more than a program. It represents a shift in AI where machines learn strategy, creativity, and flexibility in real-time." – Fei-Fei Li, Director of Stanford's AI Lab.

These achievements demonstrated AI's capacity to excel in specialized tasks, but researchers were already looking beyond specific applications. The next frontier was **generalized AI**, an intelligence system that could perform a wide range of tasks across different domains—something akin to the human ability to adapt and excel in various fields. While today's AI systems are still largely specialized, the rapid progress in learning algorithms, data availability, and computational power has brought us closer to achieving general AI than ever before.

Generalized AI continues to make significant strides, aiming to create versatile models capable of performing a wide range of tasks across different domains. Recent advancements include the development of multimodal AI systems like OpenAI's GPT-4 and Google's Gemini, which can process text, images, and videos, enhancing their utility in various applications. These models are becoming more efficient, reducing bias and errors and improving their reasoning and planning capabilities. The upgraded Ameca robot exemplifies these advancements, now displaying complex human-like emotions and more natural interactions, making it one of the most advanced humanoid robots. Additionally, Google's AI Time Machine offers groundbreaking capabilities in temporal data analysis, while OpenAI's upcoming Orion model promises to elevate AI's problem-solving abilities. Microsoft's new Copilot and autonomous agents are boosting productivity and automation, and synthetic AI humans with realistic muscles are pushing the boundaries of realism. These developments signal a future where AI reaches unprecedented levels of sophistication and impact, transforming industries and everyday life.

However, with this progress came an increasing number of ethical concerns. The rise of AI in industries like healthcare, finance, and law raised questions about accountability and fairness. How do we ensure that AI systems, which learn from historical data, do not perpetuate the biases and inequalities present in that data? Moreover, as AI begins to make decisions in critical areas—such as granting loans or diagnosing diseases—there's an urgent need for transparency in how these decisions are made.

"We need to ensure that AI serves humanity, not the other way around." – Sundar Pichai, CEO of Google, emphasizing the importance of responsible AI development.

As we move further into the 21st century, the challenge will not only be advancing AI but also ensuring that its growth aligns with ethical considerations, societal needs, and human well-being.

The Road Ahead: Challenges and Opportunities

"The advance of AI will transform every industry and create new opportunities." – Jensen Huang

As we look to the future, the road ahead for AI is filled with both incredible opportunities and daunting challenges. On the one hand, AI has the potential to solve problems that have long eluded humanity. From curing diseases to mitigating climate change, AI's ability to process vast amounts of data and uncover insights can lead to breakthroughs that will define this century. In industries such as healthcare, AI systems are being developed to assist with drug discovery and personalized medicine, enabling doctors to tailor treatments to individual patients in ways that were unimaginable a decade ago.

AI also holds the promise of transforming industries and economies. By automating mundane, repetitive tasks, AI can free up human labor for more creative and meaningful work. This shift could lead to a future where jobs that require empathy, creativity, and complex problem-solving thrive, while tasks that machines excel at are left to AI. According to a 2020 report by the World Economic Forum, AI and automation are projected to create 12 million net new jobs by 2025, transforming the workforce and the global economy.

Yet, these opportunities come with significant challenges. As AI becomes more integrated into our lives, concerns about privacy, security, and autonomy are growing. **Elon Musk**, an outspoken voice in the AI ethics debate, once remarked, *"AI is the most existential threat to humanity,"* underscoring the need for careful consideration of the paths we take in AI development. Issues such as algorithmic bias, data privacy, and the displacement of workers will need to be addressed as AI's role expands.

Moreover, there is an ongoing debate around the control and regulation of AI. Governments around the world are now grappling with how to ensure that AI develops in ways that benefit society while mitigating the risks associated with powerful, unchecked systems. In 2020, the **European Union** introduced a proposal for a legal framework to regulate AI, focusing on transparency, accountability, and the rights of individuals affected by AI systems.

"AI will not replace humans, but humans who use AI will replace those who don't." – Ginni Rometty, former CEO of IBM.

This quote reflects the evolving relationship between AI and humanity. As we march further into the age of AI, the question is not whether we will coexist with these intelligent systems but how we will ensure that AI remains an entity that empowers humanity rather than controls it. The future of AI is one we are actively shaping, and the decisions we make today will determine how AI serves us tomorrow.

Chapter 3: The Ethical Dilemmas of AI – Navigating a New Frontier

As artificial intelligence continues to permeate every corner of society, its benefits are becoming increasingly clear—efficiency, innovation, and the ability to solve complex problems are all remarkable achievements. However, along with this progress comes a myriad of ethical dilemmas that cannot be ignored. AI is not neutral—it reflects the values, biases, and decisions of the people who create it. The way we handle these ethical challenges will shape the future, determining whether AI becomes a force for good or a source of unchecked harm.

The most pressing ethical concerns surrounding AI center around **bias**, **accountability**, and **transparency**. As AI systems are trained on historical data, they often inherit the biases present in that data. In fields like law enforcement and finance, this has led to serious concerns about fairness. For example, studies have shown that AI-driven algorithms used in criminal justice systems can perpetuate racial biases, disproportionately predicting higher recidivism rates for minority groups. Similarly, in hiring processes, AI systems have been found to favor male candidates over female ones simply because the historical data reflects a gender imbalance in the workforce.

"If we're not careful, AI could reinforce and amplify existing biases, leading to more inequality rather than less." – Kate Crawford, co-founder of AI Now Institute.

Beyond bias, there's the issue of **accountability**: who is responsible when AI makes a mistake? In the case of an autonomous vehicle causing an accident, for instance, is it the fault of the manufacturer, the programmer, or the AI itself? These are not hypothetical questions. As AI begins to make decisions in high-stakes areas such as healthcare, transportation, and criminal justice, determining accountability will be crucial.

Transparency is another significant challenge. Many AI systems, especially those using deep learning, are often referred to as "black boxes" because their decision-making processes are opaque, even to their creators. This lack of transparency makes it difficult to understand how certain decisions are made, raising questions about trust and fairness. If we don't know why an AI system denied a loan application or recommended a medical procedure, how can we ensure that it's being fair or making the best decision?

These ethical dilemmas point to a larger question: as we build more powerful AI systems, how can we ensure that they serve humanity in a way that is fair, just, and transparent?

Bias in AI: Inherited Prejudices

"Artificial intelligence can be no better than the data it learns from." – Joy Buolamwini, founder of the Algorithmic Justice League.

One of the most challenging ethical issues in AI is the presence of bias. Because AI systems learn from data—data generated by humans—these systems often reflect the prejudices that exist in society. Even when programmers aim to be neutral, the data fed into AI models can encode biases related to race, gender, socioeconomic status, and more.

A well-known example of bias in AI emerged in **facial recognition technology**. Early versions of these systems had drastically lower accuracy rates when identifying people of color, particularly Black individuals and women. A study conducted by **MIT Media Lab** found that some facial recognition systems had error rates as high as 35% for Black women, compared to less than 1% for white men. This disparity led to significant public outcry, highlighting the need for AI developers to address the bias embedded in their training data.

Another notable case occurred with Amazon's AI recruitment tool. The system, designed to streamline the hiring process by automatically scanning resumes, developed a bias against female applicants. Because it had been trained on historical data that reflected male-dominated hiring practices, the AI system began penalizing resumes that contained words associated with women, such as "women's chess club." The system ultimately reinforced the gender imbalance it was supposed to help fix.

"If AI is trained on biased data, it will perpetuate those biases in its decisions and predictions." – Timnit Gebru, AI researcher and advocate for ethical AI.

These examples show how easily AI can reflect and even amplify societal inequalities. As AI systems become more ingrained in everyday life, the impact of these biases will only grow. For instance, biased algorithms in criminal justice systems have led to harsher sentencing recommendations for minority defendants, further exacerbating racial disparities in incarceration rates. AI systems used in lending can also perpetuate discriminatory practices by denying loans to certain demographic groups based on flawed data.

Addressing bias in AI is not a simple task. It requires not only improving the data we use but also implementing stronger checks and balances throughout the AI development process.

Transparency, fairness, and accountability must be built into AI systems from the ground up.

Accountability: Who Takes the Blame When AI Fails?

"When machines make decisions that affect human lives, we must be clear about who is responsible." – Virginia Dignum, AI ethics researcher.

As AI systems take on more responsibilities, the question of accountability becomes increasingly urgent. AI is now being used in high-stakes scenarios—self-driving cars, medical diagnostics, financial lending—and the consequences of AI failures can be catastrophic. But when things go wrong, who is held accountable? Is it the developers, the organizations deploying AI, or the system itself?

Two of the most high-profile examples of this dilemma occurred in October 2024; a recent incident has raised concerns about the impact of technology on mental health. A teenager tragically took his own life after developing an intense obsession with an AI chatbot. This case highlights the potential risks of excessive reliance on artificial intelligence for emotional support, especially among vulnerable individuals. It underscores the need for greater awareness and regulation of AI interactions and the importance of human connection and professional mental health support in addressing emotional and psychological challenges. In 2018, when a self-driving Uber vehicle struck and killed a pedestrian in Arizona. The incident raised complex legal and ethical questions. Should Uber, as the company behind the vehicle, be responsible? What about the engineers who programmed the AI? Or the backup driver who was in the car but not paying attention? The case illustrated the murky legal territory that exists when AI systems malfunction or make deadly mistakes.

"We've built systems that are more capable than ever, but we haven't yet built the frameworks to ensure they're used responsibly." – Elon Musk, discussing AI's rapid development.

The complexity of AI systems makes it difficult to pinpoint who—or what—is responsible when failures occur. Many AI systems, particularly those based on **deep learning**, function as "black boxes" whose inner workings are hard to interpret. Even the engineers who design these systems may not fully understand how the algorithms arrive at their decisions. This lack of transparency complicates the assignment of blame when things go wrong.

Moreover, AI failures are often systemic, resulting from multiple factors, which means no single individual or entity can be held solely accountable. For instance, if an AI system used in healthcare recommends an incorrect diagnosis, should we blame the doctor for relying on the system, the developers for creating it, or the hospital for implementing it without proper oversight?

These ambiguities underscore the need for legal frameworks that address AI accountability. In 2020, the **European Commission** proposed a set of regulations aimed at addressing AI-related risks, including transparency requirements and accountability measures. Such regulations are crucial to ensuring that, as AI takes on more responsibilities, there are clear lines of accountability to protect public safety and trust.

"If we don't have accountability, we risk creating a world where no one takes responsibility for the actions of machines." – Shannon Vallor, professor of AI ethics.

Transparency: The "Black Box" Problem

"Transparency is the foundation of trust, and without trust, AI will never fulfill its potential." – Sundar Pichai, CEO of Google.

One of the most pressing challenges in AI ethics is the issue of transparency. Many AI systems, especially those powered by deep learning, operate in ways that are difficult to explain—even to the experts who design them. This phenomenon is often referred to as the **"black box" problem**: AI makes decisions based on complex algorithms and vast amounts of data, but how it arrives at those decisions can be opaque and difficult to interpret.

For example, AI algorithms used in **credit scoring** can process hundreds of data points to determine a person's creditworthiness. While these systems can be more efficient than traditional methods, they often don't provide clear explanations for why a loan was approved or denied. This lack of transparency can have serious consequences for individuals whose financial futures depend on these decisions. How can someone appeal a decision if they don't understand why it was made in the first place?

The **black box problem** is not just an issue of fairness but also one of safety. In healthcare, AI is increasingly being used to assist with diagnoses and treatment recommendations. However, if doctors don't understand how an AI system reaches a particular conclusion, they may be reluctant to trust it. Similarly, in **autonomous driving**, engineers need to know how AI systems respond to different scenarios to ensure they can safely operate on the roads. Without transparency, it becomes difficult to identify and correct potential flaws before they lead to catastrophic outcomes.

"We cannot allow AI to make decisions in the dark. Transparency is essential to ensure these systems serve society fairly and ethically." – Cathy O'Neil, author of Weapons of Math Destruction.

Several solutions have been proposed to address the black box problem, including **explainable AI (XAI)**, a field of research focused on making AI systems more interpretable. XAI aims to create models that not only make predictions but also provide clear explanations for how they arrived at those predictions. This could be a critical step in improving trust in AI, particularly in sensitive areas like healthcare, law enforcement, and finance.

Despite these efforts, achieving full transparency remains a significant challenge. Deep learning models are inherently complex, and simplifying their decision-making processes could potentially reduce their effectiveness. Striking the right balance between transparency and performance is an ongoing struggle for AI researchers and developers.

"AI without transparency is AI without accountability." – Tim Berners-Lee, inventor of the World Wide Web.

Ethical AI: Striving for a Balance

"The ethical and social implications of AI should be at the forefront of its development." – Sundar Pichai

As AI continues to evolve, the need for ethical frameworks becomes increasingly urgent. The potential for AI to revolutionize industries and improve lives is enormous, but without proper safeguards, it could also exacerbate inequalities and lead to unintended consequences. Striking a balance between innovation and ethical responsibility is essential if AI is to serve humanity rather than harm it.

Several organizations and governments have already begun working on **AI ethics frameworks** to guide development and implementation. For example, the **European Union's General Data Protection Regulation (GDPR)** includes provisions for individuals to demand explanations when automated decisions are made about them, aiming to ensure transparency and accountability in AI systems. Similarly, companies like Google, IBM, and Microsoft have established internal AI ethics committees to review the impact of their technologies and promote ethical practices.

However, the responsibility for ethical AI cannot rest solely with governments or corporations. Developers, data scientists, and researchers all have a role to play in ensuring that the AI systems they create are aligned with values like fairness, transparency, and accountability. It's not enough to pursue technological advancement at all costs—there must be an active commitment to understanding and addressing the broader societal impact of AI.

"Ethical AI is not just a choice; it's an obligation. As creators of these systems, we are responsible for their consequences." – Timnit Gebru, AI researcher and advocate for ethics in AI.

One approach to building more ethical AI is to adopt the concept of **"human-in-the-loop"** systems. These are AI models designed to include human oversight in the decision-making process, ensuring that critical choices are reviewed and approved by a person rather than left entirely to an algorithm. This model is particularly important in fields like healthcare, law enforcement, and autonomous vehicles, where the consequences of AI decisions can be life-altering.

While ethical AI is a complex and evolving challenge, the conversation around it is critical to shaping the future of this technology. As AI becomes more integrated into the fabric of society, we must continue to question not only what AI can achieve but also how it can be developed and deployed in ways that promote the well-being of all people.

"The future of AI is in our hands. It's up to us to guide its development with responsibility, fairness, and compassion." – Satya Nadella, CEO of Microsoft.

Immediate Need for Personal Growth

As we navigate AI's ethical complexities, the responsibility doesn't rest solely with technologists and policymakers. Every individual must also adapt, developing the skills and mindset necessary to thrive in a world where AI is deeply integrated into daily life.

"The greatest gift you can give to somebody is your own personal development." – Jim Rohn.

People need to learn new skills and adapt to the changing world as AI continues to improve and become more integrated into many areas of daily life. The rise of AI brings both chances and challenges, so it's important to grow as a person to do well in this world.

1. Critical Thinking and Solving Problems: AI can do routine jobs, but people will always need to be creative and able to think critically. Learning how to think strategically and solve problems will help AI do its job better and help people make better decisions.

2. Technical literacy: Knowing the basics of AI, like how it works and what it could be used for, can help people use these technologies smartly and responsibly. This knowledge also helps people make smart decisions about how AI affects different parts of their lives and jobs.

3. Ethical Awareness: Because AI technologies bring up ethical questions and concerns, it is important to be aware of these problems and have conversations about how to use AI responsibly. To grow as a person in this area, you need to understand the ethical aspects of AI and fight for fair and equal practices.

4. Being able to adapt and learn new things throughout your life: Because technology changes so quickly, people need to be able to learn new things throughout their lives. As AI takes over the world, people will need to learn new skills and information to stay relevant and useful.

"Personal development is a necessity, not an accessory; it is the foundation upon which we build a meaningful and successful life."– Alex Salem

By focusing on personal development, we ensure that humanity continues to lead and shape its future, using AI's assistance to enrich life rather than dictate it.

As we look ahead, it's clear that personal growth and ethical awareness will be the keys to thriving in an AI-driven world. While machines excel at data analysis and predictive tasks, they lack empathy, intuition, and ethical judgment—traits that are uniquely human. By cultivating these qualities, we can ensure that we remain in control, using AI to enrich human life rather than replace it.

Chapter 4: Double-Edged Sword – The Promise and Peril of AI

"AI is a double-edged sword; without vigilance, it may carve wounds deep and wide, but with mindful hands, it can sharpen our potential, enhance humanity, and reshape our world."- Alex Salem

Artificial intelligence represents one of humanity's greatest achievements, a transformative technology with the potential to reshape countless aspects of our world. With AI, we have unprecedented power at our fingertips—an ability to address some of society's most challenging issues, from medical breakthroughs to climate change. Yet, as promising as AI is, it also presents significant risks. Like a double-edged sword, AI's power can be wielded for good or, if misused, lead to unintended and potentially harmful consequences.

This paradox is one of the most defining aspects of AI. On the one hand, it enables us to create more efficient systems, save lives, and enhance our quality of life. On the other, it introduces issues of surveillance, privacy invasion, and even job displacement, causing uncertainty about the future. The duality of AI raises critical questions about how we manage and control such a potent *tool*:

How do we maximize its benefits while minimizing its risks?

A powerful example of this dual nature is in healthcare. AI's capacity to analyze vast datasets has revolutionized diagnostics, allowing doctors to detect diseases at an early stage, personalize treatments, and improve patient outcomes. For instance, AI algorithms can now detect cancerous cells in medical scans with remarkable accuracy, potentially saving countless lives through early intervention. This is AI at its best, augmenting human skills and delivering life-changing benefits.

"With great power comes great responsibility." – Voltaire

However, the same technology that identifies disease can also be used to invade privacy. Medical AI applications require sensitive patient data, raising concerns about who has access to this data and how it might be used. In the wrong hands, personal health information could be exploited for profit or discrimination. The balance between innovation and caution is delicate, highlighting the need for careful regulation.

AI in social media: Harnessing the power, avoiding the pitfalls

"Let social media be your classroom; learn something new every day and share your insights

with the world to inspire, educate, and empower others." – Alex Salem

The rise of social media has had a profound impact on our lives. However, it was not until the dawn of AI that social media truly transformed into the powerful force it is today, both positively and negatively. On the one hand, it has revolutionized the way we connect with others, providing instant access to friends, family, and even acquaintances around the world. Social media platforms have also become invaluable resources for learning and personal growth. Online communities dedicated to specific interests, hobbies, or professions allow people to connect with like-minded individuals. The marriage of AI and social media created a potent synergy, a perfect storm that unprecedentedly reshaped the online world. AI transformed this connection into an intricate dance of algorithms and personalized content. These algorithms trained on our online behavior became experts at anticipating our needs and desires, feeding us a constant stream of content designed to keep us engaged and scrolling. The algorithms, driven by their data-driven logic, can inadvertently create echo chambers and filter bubbles, isolating us within our own pre-existing beliefs and perspectives. AI creates a feedback loop that feeds our desire for validation and connections. This feedback loop, referred to as the "social media dopamine loop," can be addictive, maximize screen time, and keep us coming back for more. The constant stream of personalized content, designed to stimulate our reward pathways and elicit emotional responses, can make it challenging to disengage from our devices and focus on other aspects of our lives.

AI in Environmental Conservation: A Powerful Ally with Risks

"AI could be one of our most powerful allies in the fight against climate change." – Demis Hassabis, CEO of DeepMind.

One of the most promising applications of AI lies in its ability to aid environmental conservation. Climate change is one of the most pressing issues facing humanity, and AI's data processing capabilities allow scientists to analyze weather patterns, predict natural disasters, and develop strategies for mitigating environmental impact. By using machine learning models to interpret vast amounts of climate data, researchers can create early warning systems that protect vulnerable populations and inform policies aimed at sustainability.

A specific example of this is AI's role in tracking endangered species. Using drones and computer vision, conservationists can monitor wildlife populations across remote regions, gathering data that would otherwise be inaccessible. These insights allow for better protection of biodiversity and help identify areas most in need of conservation efforts. In this way, AI becomes an invaluable support in the preservation of the natural world.

Yet, even in environmental applications, AI can be a double-edged sword. As AI systems rely on extensive data, there are potential privacy concerns related to the surveillance technology used to monitor protected areas. The same tools designed to safeguard ecosystems can also be leveraged for intrusive monitoring of human activity, especially in regions where governments may not always respect individual freedoms.

"We should focus on how AI can work in harmony with the natural world, not against it." – Yoshua Bengio

Furthermore, the energy requirements for powering AI models present their own environmental challenges. Training large machine learning models consumes significant electricity, often relying on non-renewable resources. The irony is clear: while AI can help combat climate change, it also contributes to it through its carbon footprint. This conflict calls for innovation not only in AI applications but also in the energy sources that support them, pushing for greener, more sustainable solutions.

"We have a responsibility to ensure AI is used responsibly, both in the protection and

preservation of our planet." – Sundar Pichai, CEO of Google.

AI in Security and Surveillance: The Line Between Safety and Privacy

"The balance between privacy and security is delicate and constantly shifting. AI adds a new layer of complexity to this equation." – Tim Cook, CEO of Apple.

One of the most controversial uses of AI lies in security and surveillance. Governments and private organizations increasingly rely on AI-powered surveillance systems to monitor public spaces, track criminal activity, and even predict potential threats before they materialize. With AI's ability to analyze massive amounts of data in real-time, security systems can quickly identify suspicious behavior, helping prevent incidents and keeping communities safer. In high-risk environments, this kind of technology can save lives, protecting citizens from threats ranging from terrorism to organized crime.

However, AI-driven surveillance also raises profound ethical concerns. Facial recognition technology, a staple in modern surveillance, has already come under scrutiny for its accuracy issues, especially when it comes to racial and gender biases. Studies have shown that facial recognition systems tend to misidentify individuals from minority groups at higher rates, which could lead to wrongful arrests, discrimination, and a loss of trust in the justice system. As a result, several cities around the world, including San Francisco, have taken steps to ban or restrict the use of facial recognition by law enforcement.

"A society that sacrifices privacy for security will lose both and deserve neither." – Benjamin Franklin.

Moreover, widespread surveillance introduces the risk of eroding individual privacy. AI-powered systems can monitor not only public behavior but also private communications and activities, blurring the line between personal freedom and governmental control. In some countries, AI-driven surveillance has been weaponized to suppress dissent and monitor citizens' every move. This potential for misuse has led to a widespread debate on the boundaries of AI in security and the ethical obligations of those who deploy it.

The challenge, then, is to ensure that AI in security serves the public interest without infringing upon personal freedoms. Strict regulations, transparency in AI deployments, and rigorous oversight can help strike a balance between safety and privacy. It's essential to remember that while AI can enhance security, it should never come at the cost of basic human rights.

"In the age of AI, we must ask ourselves: Are we protecting people, or are we controlling them?" – Edward Snowden.

AI in the Workplace: Productivity Gains vs. Job Displacement

"AI will automate tasks, not jobs. People will have to adapt by developing skills that complement AI." – Andrew Ng, Co-founder of Coursera.

One of the most widely discussed impacts of AI is its role in transforming the workplace. From streamlining operations to enhancing productivity, AI has changed the way industries operate. In sectors like manufacturing, finance, and healthcare, AI systems automate repetitive tasks, analyze complex datasets, and provide insights that would otherwise take humans weeks to uncover. These productivity gains enable companies to operate more efficiently, reduce costs, and focus human talent on more strategic and creative tasks.

For example, AI-powered customer service chatbots allow companies to provide 24/7 support, handling thousands of customer inquiries simultaneously and freeing human agents to address more complex cases. In healthcare, AI algorithms assist radiologists by identifying patterns in medical images, accelerating diagnosis, and enabling doctors to focus on patient care. In this way, AI serves as invaluable support for improving workplace efficiency and outcomes.

Yet, the rise of AI also brings concerns about job displacement. Automation of tasks that once required human labor has led to fears of widespread unemployment, particularly for roles involving repetitive or routine tasks. A report by the **World Economic Forum** estimates that by 2025, AI and automation could displace 85 million jobs globally, even as they create 97 million new ones. While these new jobs may require skills that complement AI, adapting to these changes will require a shift in workforce training and educational priorities.

"The future of work is about adapting to AI, not being replaced by it." – Ginni Rometty, former CEO of IBM.

For workers, this transition requires a commitment to continuous learning and skill development. Lifelong learning is a pillar of personal growth and should become an essential part of your life. This is a key focus of the book, which emphasizes the importance of embracing learning as a constant, ongoing process. Fields like data analysis, AI ethics, and human-AI collaboration are becoming critical as companies seek employees who can work alongside intelligent machines. The workforce of the future will need to focus on skills that AI cannot replicate easily, such as creativity, emotional intelligence, and complex problem-solving. Investing in these skills can help individuals remain competitive in an AI-enhanced job market.

"As AI takes on repetitive tasks, we must focus on what makes us uniquely human." – Satya Nadella, CEO of Microsoft.

Balancing the Double-Edged Sword: The Path Forward

"Technology should be a liberating force, not a dominating one." – Ursula Burns

As we navigate the complexities of AI's double-edged nature, the question becomes: how do we balance AI's potential benefits with its risks? This balance requires a multifaceted approach involving careful regulation, ethical consideration, and a focus on human-centered design. The future of AI depends on our ability to guide its development in ways that serve humanity and prevent harm.

One essential step is establishing strong regulatory frameworks that promote accountability, fairness, and transparency. Governments and organizations around the world are beginning to draft AI regulations to address privacy concerns, limit bias, and safeguard individual rights. The European Union, for example, has proposed a legal framework that categorizes AI applications by risk, aiming to impose stricter oversight on high-risk AI systems in fields like healthcare, law enforcement, and employment.

Public and corporate awareness of AI's ethical implications is also crucial. Companies developing

AI technology must take responsibility for implementing ethical guidelines that protect user privacy and prevent bias. Additionally, ethical AI design should prioritize transparency, allowing users and stakeholders to understand and question how decisions are made by these systems.

"The true potential of AI lies in augmenting human capabilities, not replacing them." – Demis Hassabis, CEO of DeepMind.

Ultimately, each of us has the power to shape the future by embracing lifelong learning and developing the skills needed to thrive alongside AI. By investing in personal growth—we ensure that we stay at the helm of our own destinies as technology evolves. When we nurture these uniquely human qualities, we pave the way for a future where AI becomes a force for empowerment, not domination.

AI, like any transformative technology, will reflect the values of those who create and guide it. By committing to ethical standards, transparency, and continuous personal growth, we can ensure that AI serves the greater good. It is our shared responsibility to shape this technology in ways that honor human dignity, safeguard privacy, and elevate the human experience. The future of AI is in our hands—let's shape it with purpose and integrity.

"The promise of AI lies not in what it can do for us, but in how we choose to use it." – Sundar Pichai, CEO of Google.

Chapter 5: Learning to Coexist with AI

"The relationship between human intelligence and artificial intelligence will necessarily be one of symbiosis." - Bryan Johnson

As artificial intelligence continues to integrate into almost every aspect of society, one truth becomes clear: learning to coexist with AI will be essential for both individuals and organizations. Far from being a passing trend, AI is reshaping industries, influencing daily life, and altering our understanding of what it means to work, learn, and interact. For individuals, this new reality demands adaptability, resilience, and a commitment to lifelong learning. For society as a whole, it calls for a reimagining of how we equip future generations to live in an AI-driven world.

The **future of work** is one of the most visible and immediate areas where AI's influence is being felt. Automation is transforming industries from manufacturing to finance, changing the way jobs are performed and, in many cases, reducing the need for human labor in certain tasks. As AI takes on repetitive and analytical roles, humans will increasingly focus on areas that machines cannot easily replicate, such as creativity, strategic thinking, and emotional intelligence. In this chapter, we will explore how AI is reshaping the job market, why skills like adaptability and lifelong learning are crucial, and how we can prepare the next generation for this AI-enhanced future.

AI's impact on the workforce has sparked both optimism and concern. On the one hand, it promises greater efficiency, productivity, and innovation; on the other, it raises fears of job displacement, economic inequality, and a widening skills gap. The reality is likely lying somewhere in between; while AI will transform many jobs, it will also create new opportunities for those who are prepared to embrace change and evolve alongside it. The future of work will demand individuals who are adaptable, proactive, and skilled in areas where AI falls short.

The Future of Work: How AI Will Reshape Industries and Job Markets

"In the future, there will be two kinds of people: those who tell computers what to do and those who are told by computers." – Marc Andreessen, a technology entrepreneur.

AI's ability to automate tasks and analyze massive datasets is transforming industries across the board. In sectors like manufacturing, finance, healthcare, and customer service, AI-driven systems are performing routine tasks with speed and accuracy, freeing up human workers for more strategic and creative roles. This shift presents both opportunities and challenges. While companies benefit from increased productivity and efficiency, workers face the need to adapt to changing job requirements or risk obsolescence.

For example, in the **manufacturing industry**, automation has streamlined assembly lines, enabling faster production with fewer errors. Robots can now perform tasks like welding, packaging, and quality control with unparalleled precision, reducing labor costs and increasing output. While this innovation drives progress, it also raises concerns about the future of traditional manufacturing jobs. Yet, new roles are emerging in areas like robotics maintenance, AI supervision, and system analytics—positions that require advanced skills and continuous learning.

In **finance**, AI is transforming everything from stock trading to fraud detection. Algorithms can analyze market trends in real-time, executing trades based on data patterns that would be impossible for humans to process manually. AI systems also identify potential security breaches, detecting irregularities in transaction data to prevent financial fraud. But as AI takes on these roles, financial analysts and traders are evolving their roles, focusing on high-level strategy, regulatory compliance, and personalized client services—tasks where human insight remains crucial.

"AI will transform many jobs, but it will also create new ones. Our job is to prepare for the shifts that lie ahead." – Ginni Rometty, former CEO of IBM.

While AI's impact varies across industries, a common thread is a growing emphasis on **lifelong learning**. With technology changing rapidly, workers will need to continually acquire new skills to remain competitive. This shift requires a mindset focused on adaptability, resilience, and self-directed learning. Embracing lifelong learning ensures that individuals remain valuable contributors to the workforce, even as AI continues to reshape the job landscape.

Compression of AI and Human Skills: The Irreplaceable Human Edge

"The art challenges the technology, and the technology inspires the art." – John Lasseter, filmmaker and animator.

As AI grows more capable, there are certain skills that it still cannot replicate—skills that define what it means to be human. While machines excel in tasks that involve data processing, pattern recognition, and automation, they fall short in areas that require **creativity**, **emotional intelligence**, **empathy**, and **complex problem-solving**. These uniquely human skills are likely to become increasingly valuable as AI takes on more routine and analytical roles.

Creativity is one area where human ingenuity remains unmatched. While AI can generate text, art, and music by analyzing patterns, it lacks the true spontaneity, imagination, and intentionality that drive human creativity. AI can suggest ideas or generate drafts, but it cannot bring the same depth of personal experience or intuition that a human artist, writer, or musician would bring. This skill is crucial in fields like advertising, design, and entertainment, where the ability to tell compelling stories and engage audiences on an emotional level is essential.

Similarly, **emotional intelligence** and **empathy** are qualities that AI cannot emulate. In professions like healthcare, counseling, and customer service, connecting with others on a human level is paramount. While AI-powered chatbots can handle basic inquiries, they lack the ability to respond to nuanced emotions or recognize the underlying needs of a person in distress. This limitation underscores the importance of human workers who can listen, empathize, and provide comfort and understanding.

"AI can provide insights, but it is up to humans to interpret and act on them." – Kai-Fu Lee

Problem-solving and **critical thinking** also remain distinctively human skills. In complex, unpredictable situations, humans can adapt, weigh ethical considerations, and make decisions based on context and intuition—factors that are difficult to program into AI systems. This skill set is indispensable in areas like law, project management, and emergency response, where the stakes are high and there is no one-size-fits-all solution.

As we move into an AI-driven future, these human skills will not only coexist with AI but will complement it. People who can harness the strengths of AI while bringing their own unique talents will become indispensable in a rapidly evolving workforce.

Preparing the Next Generation: Teaching Children and Young People to Thrive with AI

"Education is the most powerful weapon which you can use to change the world." – Nelson Mandela.

Preparing this generation and future generations to coexist with AI is a vital aspect of ensuring a balanced, productive, and ethically aware society. As AI increasingly shapes the world of work and daily life, equipping young people with the right skills, knowledge, and mindset becomes critical. This preparation involves not only technical training but also nurturing **resilience**, **ethical awareness**, personal development, and a capacity for **lifelong learning**.

One essential aspect of preparing the next generation is fostering **technical literacy** from an early age. Today's children and young adults will grow up in a world where AI is embedded in nearly every industry, and understanding the basics of AI, machine learning, and data processing will empower them to navigate this landscape effectively. Programs in schools, community centers, seminars, and online classes can introduce concepts of AI in accessible ways, from hands-on coding workshops to lessons about how algorithms affect everything from social media to healthcare. By demystifying AI, we give young people the tools to use these technologies responsibly.

Beyond technical skills, **ethical education** around AI is crucial. As future developers, users, and decision-makers, today's youth need a solid foundation in the ethics of technology. They must understand the implications of bias, privacy, and accountability in AI, learning to question how these systems impact individuals and communities. For instance, discussing real-world cases of AI bias in facial recognition or the ethical dilemmas of AI-driven surveillance can instill a sense of responsibility and awareness. This grounding in ethics empowers young people to approach AI with an understanding of both its capabilities and its limitations.

"Teaching children to navigate AI isn't just about tech skills; it's about fostering critical thinking and ethical responsibility." – Tim Cook, CEO of Apple.

Additionally, promoting **creativity and adaptability** will prepare the next generation for a future where AI performs many routine tasks. Skills like creativity, critical thinking, and problem-solving—qualities that AI struggles to replicate—will become increasingly valuable in the job market. Educational programs can encourage these qualities through project-based learning, teamwork, and problem-solving exercises, helping young people build the versatility they'll need to thrive alongside AI.

By embedding these skills and mindsets into education and committing to lifelong learning, we empower future generations not just to work with AI but to shape it in ways that serve humanity.

Embracing Lifelong Learning: Adapting in an Ever-Changing World

"The illiterate of the 21st century will not be those who cannot read and write, but those who cannot learn, unlearn, and relearn." – Alvin Toffler, futurist and author.

As AI continues to evolve, the concept of **lifelong learning** has become a cornerstone for personal and professional growth. The rapid pace of technological advancement means that skills and knowledge that are valuable today may quickly become outdated. To remain adaptable and relevant in a world where AI automates and accelerates many tasks, individuals must cultivate a mindset of continuous improvement, self-development, and self-directed learning.

Lifelong learning goes beyond formal education; it's about staying curious and proactive, seeking out new skills, and adapting to emerging trends. For instance, as AI technology reshapes industries, many people are pursuing certifications, online courses, and workshops to update their skill sets. By embracing platforms like Coursera, LinkedIn Learning, Khan Academy, and other social media platforms, individuals can acquire new competencies in everything from data analysis to AI ethics, often at their own pace and convenience.

Another key aspect of lifelong learning is **developing soft skills**—qualities like adaptability, critical thinking, and effective communication—that remain essential regardless of technical trends. These skills enable individuals to work alongside AI in dynamic environments, solve complex problems creatively, and connect with others in meaningful ways. While AI can process data, it cannot replicate the human ability to adapt quickly to new challenges or understand the emotional nuances of communication.

"In a world where AI evolves every day, we must embrace lifelong learning to keep pace. Commit to growth, adapt to change, and empower yourself to thrive in an ever-changing future."- Alex Salem

As AI changes the demands of the job market, individuals who approach learning as a lifelong commitment will be better positioned to navigate these shifts and seize new opportunities. This adaptability will be key not only for personal success but also for societal resilience. In a world increasingly shaped by AI, the ability to learn, unlearn, and relearn will enable us to coexist with intelligent machines while retaining our distinctively human strengths.

Ultimately, learning to coexist with AI isn't just about adopting new skills; it's about cultivating a growth mindset that allows us to evolve in harmony with this transformative technology. By embracing lifelong learning, nurturing essential human skills, and preparing future generations, we create a future where AI and humanity progress together, each enriching the other.

Chapter 6: AI is an Instant Solution, Not a Replacement

Enhancing, Not Replacing: The Unique Value of Human Abilities

"AI can calculate, but only humans can care." – Anonymous

Artificial intelligence has made remarkable strides in processing information, solving problems, and even mimicking aspects of human creativity. However, it remains fundamentally different from human intelligence. While AI excels in efficiency, speed, and data-driven decision-making, it lacks the innate qualities that define human intelligence—**creativity, intuition, and empathy**. These are the traits that inspire art, fuel scientific discovery, and connect us to one another in meaningful ways.

In fields like healthcare, AI assists doctors in diagnosing diseases, but it cannot replace the empathy and understanding a doctor offers when delivering a difficult diagnosis. Similarly, in creative industries, AI-generated art or music might impress technically, but it rarely carries the depth of emotion or personal insight of a human creator. This distinction reflects AI's role as a ***tool*** **to**

enhance human abilities, not a substitute for the human experience.

Human creativity, for example, is fueled by individual experiences, cultural context, and a sense of purpose—all of which are unique to humans. An AI system can generate an endless array of paintings based on data, but it cannot create with intention or personal narrative. Likewise, intuition—a critical asset in professions like psychology, counseling, and leadership—relies on an understanding of subtle social cues and emotional resonance, which AI simply cannot replicate.

"AI is here to augment our abilities, not to redefine our humanity." – Satya Nadella, CEO of Microsoft

As we explore the potential of AI, we must remember that it's not about replacing human intelligence. It's about amplifying our strengths, providing us with insights that help us make better, faster decisions, and enabling us to approach problems in ways we could not achieve alone. AI is an instant solution—a powerful *tool* to complement our efforts—yet it does not diminish the fundamental qualities that make us uniquely human.

Collaboration, Not Competition: AI and Human Partnerships in Industry

"Alone we can do so little; together we can do so much." – Helen Keller

The relationship between AI and human capabilities is most effective when viewed as a **collaborative partnership** rather than a competitive dynamic. Industries around the world are discovering that when AI and humans work together, they can achieve results that neither can accomplish alone. From healthcare and education to entertainment and business, the collaboration between humans and AI allows for the best of both worlds: the precision and efficiency of technology coupled with the creativity, intuition, and empathy unique to humans.

In **healthcare**, for example, AI assists doctors in analyzing patient data, identifying potential health risks, and developing treatment plans tailored to individual needs. However, it's the human doctors who apply critical thinking, experience, and empathy to interpret these insights and provide compassionate care. This collaboration not only improves diagnosis and treatment accuracy but also ensures that patients feel understood and supported—a balance only achievable when AI and human skills complement each other.

In **education**, AI can adaptively tailor learning materials to individual students' needs, recognizing patterns in their learning styles and challenges. Yet, the role of the human educator remains irreplaceable. Teachers offer mentorship, motivation, and interpersonal connections that foster a positive learning environment. By working together, AI and educators create an environment where each student receives personalized support while benefiting from the encouragement and guidance of a caring mentor.

"Human creativity and AI's precision together can achieve epic results." – Alex Salem

In the **entertainment industry**, AI is used to analyze audience preferences and suggest trends, but it's the human creators who craft stories, develop characters, and produce content that resonates on an emotional level. The insight provided by AI allows creators to understand their audiences better, but it's the human touch that ultimately brings entertainment to life. By collaborating rather than competing, AI and humans can enhance the quality, relevance, and reach of creative work.

Empowering Humans Through AI: Stories of Innovation and Positive Change

"Technology is best when it brings people together." – Matt Mullenweg, Co-founder of WordPress

The power of AI lies not only in its technical capabilities but also in its potential to **empower individuals and communities**. Around the world, people are harnessing AI to create positive change, solve complex problems, and innovate in once unimaginable ways. These stories illustrate how AI, when used thoughtfully, can be a powerful force for good.

One such story comes from **farmers in rural India** who have used AI-driven platforms to gain critical insights into crop health, weather predictions, and market trends. With limited access to traditional agricultural resources, these farmers can now monitor crop conditions through smartphone apps, receive real-time advice on pest management, and access fair pricing information for their produce. Through AI, these communities are not only improving their yield and profitability but also becoming more resilient to environmental and economic changes.

Another powerful example is **healthcare accessibility**. In regions with limited access to medical professionals, AI-powered diagnostic tools are bridging the gap by enabling basic health assessments remotely. For instance, AI-driven mobile applications are helping people detect skin conditions, eye diseases, and other health issues through simple scans, often guiding them to seek medical help when needed. These tools empower individuals to take charge of their health, even in underserved areas, reducing barriers to care.

"The impact of AI depends on how we design it and the purposes we choose to put it to." – Kate Crawford

In education, AI is helping to break down barriers for students with disabilities. Speech recognition and natural language processing technologies allow students with hearing or speech impairments to engage with learning materials through adaptive interfaces. This technology empowers students to participate fully in classrooms and access educational resources in previously inaccessible ways, fostering a more inclusive learning environment.

These stories highlight a central paradox: AI empowers people, enabling them to overcome limitations, access resources, and build stronger communities. But as it amplifies our ability to effect change and make a positive impact, it also raises a question—does AI merely serve us, or is it beginning to shape the very nature of our agency?

Collaboration Over Competition: A Vision for the Future

"The best way to predict the future is to create it." – Peter Drucker, management consultant and author

As we look toward the future, it's evident that AI and human skills are not on opposing sides. Instead, they represent two halves of a greater potential when combined. A future where AI and humans work in harmony can elevate every field—from healthcare and education to creative arts and customer service. Rather than fearing replacement by AI, individuals and industries can embrace **collaborative innovation**, where AI takes on tasks that it performs well, freeing people to focus on the aspects of work that require human insight and emotional connection.

In **business**, the collaboration between AI and humans can lead to new insights and improved strategies. Companies are increasingly adopting AI for data analysis and customer behavior prediction, yet it's human teams that interpret these insights, make strategic decisions, and build relationships with clients. By working together, AI helps businesses anticipate trends and respond more effectively, while human insight drives creativity, trust, and long-term value.

In **creative industries**, where imagination and originality are paramount, AI can provide inspiration, suggest new ideas, or even enhance creative techniques. Musicians, for example, are using AI tools to explore new compositions, blending styles, or experimenting with sounds that push traditional boundaries. However, it is the musician's personal experience, intuition, and creative vision that turn AI-generated suggestions into a final piece of art that resonates emotionally. Rather than replacing artists, AI acts as a partner that broadens the possibilities of creative expression.

"In the age of AI, humans and machines will do what they each do best. We'll let AI handle the data, and humans handle the meaning." – Garry Kasparov, chess grandmaster and AI advocate

The **education sector** further exemplifies how collaboration can drive progress. AI supports teachers by providing personalized lesson plans and identifying areas where students may need extra support. However, teachers provide mentorship, motivation, and guidance—factors essential to a student's growth that AI cannot replace. Together, AI and educators create a learning experience that is both efficient and deeply meaningful.

This vision of collaboration over competition offers a sustainable and empowering future. By working alongside AI, we leverage its strengths while amplifying our unique human qualities, fostering a world where technology supports rather than overshadows humanity.

Chapter 7: AI and the Supremacy of Choice

The Importance of Mindful Use: Exercising Control in the Age of AI

"With the rise of AI, our choices define our future. Let us wield this power with wisdom and responsibility." – Alex Salem

As AI becomes deeply embedded in our lives, we are presented with an unprecedented level of convenience and efficiency. AI can streamline tasks, predict our needs, and even enhance our decision-making processes. But with this immense power comes the critical responsibility to make **mindful choices** about how we engage with this technology. Mindful use of AI requires us to remain conscious of both the benefits and the potential consequences, understanding that every interaction with AI shapes our relationship with it.

Being mindful involves asking ourselves essential questions: Are we using AI to enhance our capabilities or to delegate responsibilities we'd be better off handling ourselves? Are we aware of how much personal data we're sharing with these systems, and do we understand how this data might be used or shared in ways that could affect us in the long run? These questions may seem small at the moment, but they define the nature of our relationship with AI—one where we either remain in control or risk becoming dependent.

One example of mindful use is in **personal finance**. Many people now rely on AI-powered budgeting apps, stock-trading algorithms, and automated saving tools. These applications offer immense convenience, helping users track expenses, make investments, and set financial goals. However, while these tools are useful, there's a risk that we may start depending on them entirely, potentially losing our understanding of personal finances. By making intentional decisions—like periodically reviewing our finances independently or understanding the principles behind automated suggestions—we ensure that AI remains a *tool* rather than a crutch.

"The first principle is that you must not fool yourself—and you are the easiest person to fool." – Richard Feynman, physicist

The mindful use of AI is a safeguard against complacency. It allows us to actively shape our experiences with AI, choosing which areas to automate and where to stay hands-on. In doing so, we create a balance where AI enhances our lives without undermining our autonomy, ensuring that we remain the ultimate decision-makers.

Setting Boundaries: Regulating AI to Ensure It Works for Us, Not Against Us

"It is not enough to be industrious; so are the ants. What are you industrious about?" – Henry David Thoreau

As AI becomes more integrated into our daily lives, it's essential to establish **boundaries** that prevent it from overstepping into areas that should remain within human control. Setting boundaries is not about resisting progress; rather, it's about ensuring that AI serves us in ways that align with our values, privacy, and autonomy. With clear boundaries, AI remains a *tool* that amplifies our abilities rather than a force that dictates our behaviors.

Boundaries in AI usage can take many forms. **Regulatory frameworks** set by governments and organizations play a critical role in defining what AI can and cannot do. For example, data privacy laws like the General Data Protection Regulation (GDPR) in the European Union impose strict rules on how companies collect, store, and share user data, providing individuals with greater control over their information. Similarly, proposed regulations in the United States and other countries aim to restrict the use of AI in sensitive areas like facial recognition and autonomous decision-making, establishing boundaries that protect citizens' rights.

"Harness self-discipline by setting time limits and disabling notifications to master online interactions, curb addiction, and manage AI's impact on your time." - Alex Salem

On a personal level, setting boundaries involves limiting AI's role in certain aspects of life. For instance, social media platforms use AI algorithms to recommend content, encouraging prolonged engagement. By setting personal boundaries, such as time limits on these platforms or choosing to disable certain notifications, users can take control over their online interactions, ensuring that AI doesn't consume more of their time than intended. This practice empowers individuals to remain aware of their habits and maintain a healthy relationship with technology.

"The goal is to keep the relationship between human and machine symbiotic, not parasitic." – Jaron Lanier, computer scientist and VR pioneer

Organizations, too, can set boundaries for AI within the workplace by prioritizing transparency and oversight in AI applications. For example, companies using AI for employee performance assessments should ensure that human managers review and interpret the results rather than relying solely on AI-generated metrics. This approach ensures that while AI contributes valuable insights, human judgment and empathy remain at the forefront of decision-making.

In setting boundaries, we acknowledge that AI has limitations and that there are areas where human qualities—judgment, ethics, empathy—are irreplaceable. By defining these boundaries, we keep AI's role in check, ensuring that it serves humanity rather than controls it.

Balancing Convenience and Privacy: Navigating the Benefits and Risks

"The cost of convenience is often our privacy." – Shoshana Zuboff, author of The Age of Surveillance Capitalism

One of the most complex challenges in the age of AI is finding a balance between the convenience that AI offers and the privacy risks it brings. AI-powered devices and services are designed to make our lives easier—personal assistants help manage our schedules, smart home systems respond to our needs, and predictive algorithms suggest content tailored to our preferences. Yet, each convenience often comes at the cost of **data privacy** and personal security.

Every time we interact with AI-driven systems, whether it's asking a smart assistant for directions or using a fitness app that tracks our health metrics, we're generating data. This data, while invaluable for improving the accuracy and usefulness of AI, also holds personal details that can reveal much about our habits, behaviors, and even our innermost thoughts. Without mindful engagement and boundaries, we risk giving away our privacy in exchange for the ease that AI offers.

An example of this trade-off can be seen with **smart home devices**. Smart speakers, thermostats, and cameras enhance convenience, allowing us to control our environments with a simple voice command or automated settings. However, these devices often come with surveillance features, and they're capable of recording conversations or tracking our daily routines. Such data is valuable to companies looking to refine their products or target ads, but it also poses a privacy risk if mishandled or exposed.

"If you're not paying for the product, you are the product." – Douglas Rushkoff, media theorist

This quote highlights an essential reality: many AI-driven services are offered "for free" at the expense of user privacy. Social media platforms, for instance, use AI to deliver highly personalized feeds that keep users engaged, but this personalization is often based on extensive data collection. As we navigate this landscape, it's vital to make informed choices about which data we're willing to share and set privacy limits where possible.

By consciously evaluating where convenience is worth the potential privacy risks, individuals can create a balanced approach to AI use. This might mean disabling certain data-sharing permissions, using encryption tools, or choosing AI services that prioritize user privacy. Balancing convenience and privacy requires an ongoing assessment of our priorities, ensuring that we reap the benefits of AI without sacrificing control over our personal information.

The Power of Informed Decisions: Choosing How We Engage with AI

"It's not hard to make decisions when you know what your values are." – Roy Disney

As AI continues to weave itself into the fabric of our lives, the power to make **informed, intentional choices** becomes more significant. The ways we engage with AI—from which apps we use to the data we allow them to collect—ultimately shape our experience with technology and its impact on our lives. Understanding AI's strengths and limitations enables us to approach these interactions thoughtfully, making decisions that align with our values and priorities.

Informed decision-making means understanding both **the capabilities and potential risks** of AI systems. For instance, predictive algorithms in social media can enhance our experience by recommending content that aligns with our interests, but they can also limit our perspective by creating echo chambers that reinforce our views. By choosing to diversify our sources of information or seeking content beyond algorithmic suggestions, we exercise agency over our exposure to varied viewpoints, avoiding a filtered reality that restricts understanding.

In the realm of **personal finance**, individuals can make informed choices by understanding the logic behind AI-driven investment platforms. Robo-advisors, for example, offer tailored investment options based on an individual's financial goals, risk tolerance, and time horizon. However, users need to grasp how these algorithms operate, as relying solely on an AI recommendation can result in overconfidence or a lack of oversight. Deciding to review investments periodically and understand the reasoning behind algorithmic suggestions fosters an informed approach to financial planning.

"Choice is the most powerful tool we have. Everything boils down to the choices we make." – Oprah Winfrey

The responsibility to make conscious choices also applies to data privacy. By understanding the potential uses of our data, we can decide which permissions to grant and which to deny, giving us control over how AI interacts with our personal information. This may involve actively managing app permissions, avoiding services that overreach in data collection, or using privacy-oriented tools that minimize data exposure. Taking charge of these decisions allows us to embrace the benefits of AI without compromising personal boundaries.

Informed decision-making is also a cornerstone of personal growth and development. It requires critical thinking, self-awareness, and emotional intelligence—key elements for dealing with life's challenges, setting and achieving meaningful goals, and making choices that truly reflect your values and aspirations.

Engaging thoughtfully with AI ensures that technology enhances our lives on our own terms. It's not about avoiding AI but about making conscious and deliberate choices that shape technology's role in our lives rather than letting it dictate our actions.

PART II:
Creating Yourself & Personal Growth

Chapter 8: What is Personal Growth?

Defining Personal Growth: A Journey of Self-Discovery and Transformation

"Personal growth is not a destination; it's a journey of becoming more of who you are." – Jim Rohn

Personal growth is the ongoing process of understanding oneself, embracing change, and striving for improvement. It is a journey that encourages us to reflect on who we are and who we wish to become. Unlike achievements that can be measured by external standards, personal growth is deeply personal and inward, driven by our unique values, aspirations, and experiences. It involves a willingness to confront both strengths and weaknesses, to learn continuously, and to shape our identity in ways that align with our core beliefs.

"Think of your mind, body, and soul as teammates in a group project, striving together to get the highest, best level, and best score; they should be doing their best work and ensure that the collective goal is attained." – Alex Salem

At its essence, personal growth is about **self-creation, not self-finding.** It is the intentional journey of developing the *mind*, *body*, and *spirit*, setting the foundation for a more fulfilling life. Whether this journey involves enhancing our emotional intelligence, developing new skills, or redefining our goals, personal growth empowers us to create a life that feels authentic and meaningful. It encourages us to look inward, challenging us to grow from our past and live intentionally in the present.

As we embark on this journey, **reflection** becomes a crucial tool. Reflecting on our experiences helps us understand how they've shaped our thoughts, behaviors, and beliefs. It offers the clarity needed to make conscious decisions that align with our evolving identity. Reflection allows us to identify areas for improvement and recognize the growth we've already achieved, reminding us that change is possible and within reach. Reflecting on personal growth and development is a powerful way to understand yourself better and foster continuous improvement.

Here are some steps to help you reflect effectively:

Find a Quiet Space: Choose a comfortable, distraction-free environment where you can focus.

Set Aside Time: Dedicate specific times for reflection—daily, weekly, monthly, or yearly—to build consistency.

Ask Deep Questions: Reflect with questions like "What did I learn today?" or "How did I handle allenges?” to gain insights into your actions and cisions.

Honest with Yourself: Recognize both your engths and areas for growth without judgment.

rite It Down: Keep a journal to clarify your ughts, track your progress, and identify patterns er time.

ek Feedback: Engage with trusted friends, entors, or colleagues to gain new perspectives and sights.

t Goals: Use your reflections to create actionable als that align with your values and aspirations.

Review and Adjust: Regularly revisit your reflections and goals, making adjustments to stay on track and adapt to change.

Adopting these practices in your daily life can enhance your self-awareness, improve decision-making, and make your personal growth journey meaningful and effective.

"Growth is painful. Change is painful. But nothing is as painful as staying stuck somewhere you don't belong." – Mandy Hale

Personal growth requires us to confront discomfort and get comfortable with the uncomfortable, as true change often lies outside our comfort zone. The process of learning, unlearning, relearning, and adapting to new insights or challenges builds resilience. In this way, personal growth is not just about achieving success or happiness but about cultivating the strength to navigate life's inevitable ups and downs with wisdom and grace.

Self-Awareness: The Foundation of Personal Growth

"Knowing yourself is the beginning of all wisdom." – Aristotle

Self-awareness is the cornerstone of personal growth. It is the ability to look inward, to understand our thoughts, emotions, and motivations, and to recognize how they shape our actions and interactions. Self-awareness invites us to question why we make certain choices, why we react the way we do, and what truly matters to us. By understanding ourselves on a deeper level, we gain clarity and the ability to steer our lives in meaningful directions.

Building self-awareness requires consistent reflection. Taking time to assess our emotions, beliefs, and behaviors helps us identify patterns and uncover truths that we might otherwise overlook. Reflection can be as simple as journaling or as structured as setting aside time each week to evaluate recent experiences. As we grow more self-aware, we become better at identifying what aligns with our values and what does not, empowering us to make choices that bring us closer to our ideal selves.

Self-awareness also enhances our relationships. By understanding our own strengths and weaknesses, we can communicate more openly, empathize more deeply, and build connections that are based on authenticity. Self-aware individuals are better equipped to set healthy boundaries, acknowledge the impact of their words and actions on others, and cultivate relationships that are mutually supportive and fulfilling.

"The more you know yourself, the more clarity there is. Self-knowledge has no end." – Jiddu Krishnamurti

Through self-awareness, we gain the courage to confront the parts of ourselves that need growth and improvement. This courage propels us forward, helping us embrace change and recognize that our potential is not limited by our current circumstances. Self-awareness is both a guiding light and a powerful motivator in the journey of personal growth, setting us on a path to live authentically and with intention.

Constant Learning: Embracing a Growth Mindset

"Live as if you were to die tomorrow. Learn as if you were to live forever." – Mahatma Gandhi

A critical component of personal growth is the commitment to **constant learning**. Embracing a growth mindset—the belief that abilities and intelligence can be developed through dedication and effort—allows us to see life as a journey of continuous improvement. Rather than viewing challenges as obstacles, a growth mindset encourages us to see them as opportunities for learning and transformation.

Constant learning doesn't mean formal education alone; standard education results in standard living. We must pursue continuous learning beyond traditional education. While formal education provides the foundation for a stable career, self-education allows individuals to unlock their full potential and achieve greater success. It encompasses a wide range of experiences, from reading books, listening to audio content, watching positive educational shows, and attending workshops to engaging in new hobbies and exploring unfamiliar perspectives. By remaining open to learning, we keep our minds flexible and responsive, able to adapt to new information and evolving circumstances. This openness is essential in a world that is constantly changing, where staying stagnant can lead to missed opportunities and a feeling of disconnection.

One key to lifelong learning is **curiosity**. A curious mind seeks to understand the world more deeply, asking questions that lead to greater insight and awareness. By embracing curiosity, we allow ourselves to break free from routine thinking, enabling us to grow in ways we may not have imagined. Curiosity fuels our motivation to learn and helps us push past limitations, keeping our minds sharp and engaged with life.

Here are some benefits of curiosity and learning, which sharpen the mind and stimulate brain cells:

Creating New Connections: Learning forms new neural pathways.

Generating New Neurons: Activities like learning and exercising promote new brain cells.

Strengthening Synapses: Learning improves communication between neurons.

Speeding Up Signals: Myelin production increases, making brain signals faster.

Releasing Dopamine: Learning releases dopamine, making it enjoyable and reinforcing the process.

"Intellectual growth should commence at birth and cease only at death." – Albert Einstein

Constant learning also builds resilience, preparing us for life's uncertainties. When we cultivate a habit of learning, we become more adaptable, better equipped to handle challenges, and more confident in our ability to overcome setbacks. This resilience is one of the most powerful outcomes of personal growth, as it allows us to maintain our momentum, even when faced with adversity. By committing to constant learning, we transform our approach to life, viewing each moment as an opportunity to become a better, wiser version of ourselves.

Constant learning not only enhances our knowledge and skills but also strengthens our emotional and mental resilience, helping us navigate life's inevitable uncertainties. Life is unpredictable, and challenges—whether personal, professional, or emotional—are a given. However, when we actively engage in the habit of continuous learning, we prepare ourselves to face these challenges with a stronger, more adaptable mindset.

Learning doesn't just equip us with facts or technical skills; it builds a kind of mental flexibility. This adaptability means that when obstacles arise—whether they come in the form of failure, loss, or unexpected change—we don't easily falter. Instead, we approach these difficulties with a mindset that is focused on growth rather than defeat. Through learning, we develop problem-solving skills, creativity, and a deeper understanding of how to manage and mitigate stress.

Furthermore, the confidence gained from acquiring new knowledge and skills feeds into our resilience. As we expand our knowledge, we reinforce our belief in our ability to adapt, improve, and ultimately thrive, even in tough situations. This sense of self-efficacy—that belief that we can handle whatever life throws at us—is one of the most powerful outcomes of personal growth through learning.

This resilience enables us to maintain momentum even in the face of adversity. It ensures that we don't get stuck in moments of failure or setback but instead see them as opportunities to learn, reflect, and rise stronger. Each lesson, each challenge, becomes a stepping stone on our path to becoming a more capable, wiser version of ourselves.

By committing to lifelong learning, we transform the way we approach life itself. We no longer see obstacles as insurmountable but as valuable lessons that shape our character and abilities. This mindset shift allows us to embrace uncertainty and change, knowing that with every new experience, we are becoming more resilient, more knowledgeable, and more prepared for whatever lies ahead.

"Don't fear trying new things. Embrace the unknown with courage and curiosity, for it is in the uncharted territories that we uncover our true potential and achieve extraordinary greatness." - Howard Lorber, American Businessman, Investor, and my Mentor.

Chapter 9: Defining Your Core Values

Understanding Core Values: The Foundation of a Fulfilling Life

"Values are like fingerprints. Nobody's are the same, but you leave them all over everything you do." – Elvis Presley

Core values are the fundamental principles that guide our actions, decisions, and interactions with others. They are the bedrock of our identity, influencing how we view the world, relate to people, and pursue our goals. Core values represent what we stand for and what truly matters to us, shaping our choices and defining our sense of self. While they may evolve over time, our core values remain a consistent force, anchoring us through life's inevitable changes and challenges.

Distinguishing **core values from core beliefs** is essential to understanding their role in personal growth. Core beliefs are often shaped by our upbringing, culture, and experiences. They are assumptions we hold about ourselves and the world, but they may or may not align with our authentic values. Core values, on the other hand, are conscious, deeply held principles that we choose to uphold because they resonate with our true selves. While core beliefs may be questioned or revised, core values are intentionally embraced as guiding principles.

Core values provide a **sense of direction** and purpose. When we align our actions with our values, we feel a sense of integrity and fulfillment. Conversely, when we act against our values, we often experience discomfort, regret, or dissatisfaction. Recognizing our core values empowers us to live in a way that feels authentic, helping us to make choices that are consistent with who we are at our core.

" If you don't stick to your values when they're being tested, they're not values: they're hobbies. " – Jon Stewart

This chapter will explore the significance of core values in personal growth, guiding readers through a process of introspection to uncover their own values. By identifying and prioritizing what matters most, readers can gain clarity on their life's purpose, enhance decision-making, and build a strong foundation for a life of integrity and alignment.

Why Core Values Matter: Living a Life Aligned with Your True Self

"Embrace your core values, for they are the compass that guides you to your true self." – Alex Salem

Understanding and defining our core values is essential for personal growth because these values act as a compass, guiding our choices and helping us stay true to ourselves. When we know what we truly stand for, we become more resilient in the face of challenges and more deliberate in our decision-making. Core values provide us with a foundation of self-respect, self-love, ethics, integrity, and inner peace, offering clarity and consistency amidst life's complexities. They guide our decisions, shape our behaviors, and help us stay true to ourselves.

Living according to our values creates a sense of **inner harmony**. For example, someone who values honesty will feel most at ease when they are truthful, even in difficult situations. By aligning our actions with our values, we reduce internal conflicts and enhance our sense of self-respect. When we act against our values, however, we may experience regret or dissonance, as if we are betraying our true selves.

Core values also help us **navigate relationships**. By understanding what matters to us, we can connect with others who share similar principles, building relationships that are meaningful and fulfilling. In contrast, relationships that conflict with our values may feel forced or unfulfilling. Knowing our core values helps us set boundaries and cultivate connections that align with who we are.

"When your values are clear to you, making decisions becomes easier." – Roy T. Bennett

In times of uncertainty or stress, our core values serve as an anchor, reminding us of what truly matters. They help us make choices that are in line with our goals and aspirations rather than reacting impulsively or based on external pressures. When we are faced with difficult decisions, reflecting on our core values can bring clarity and confidence, guiding us toward paths that support our personal growth and sense of purpose.

Distinguishing Core Values from Core Beliefs: Understanding the Difference

"Your beliefs don't make you a better person; your behavior does." – Anonymous

While core values and core beliefs are often interconnected, they serve different purposes in our lives and can sometimes even conflict. Core beliefs are assumptions and perceptions that we form over time, influenced by culture, upbringing, and personal experiences. They shape our worldview, our sense of identity, and our judgments. However, some core beliefs might not align with who we are deep down; they may reflect societal expectations or inherited patterns rather than our authentic selves.

Core values, on the other hand, are **intentional choices** about what we stand for and what principles we hold closest to our hearts. Values like integrity, compassion, freedom, dedication, discipline, dignity, and growth reflect what we consciously choose to prioritize. While core beliefs can be revised as we gain new insights, core values remain more stable, acting as a foundation upon which we build our lives.

For example, someone might hold a core belief that "success means financial wealth" due to societal conditioning, but their core value might prioritize "balance and well-being" over financial gain. Recognizing the difference helps in make choices that resonate more deeply with our true selves rather than following beliefs that may not fully serve us.

"If you don't stand for something, you will fall for anything." – Malcolm X

Exercises to distinguish values from beliefs can be highly illuminating. One approach is to reflect on specific life decisions and consider what guided them: Was it a core belief based on past experiences, or was it a value reflecting what you genuinely prioritize? By examining these distinctions, we gain clarity about what drives our actions and how to realign ourselves with values that bring us closer to the life we aspire to lead.

Identifying and Prioritizing Your Core Values: Exercises for Self-Discovery

"The unexamined life is not worth living." – Socrates

Discovering your core values requires self-reflection and a willingness to examine what truly matters to you beyond societal expectations or the influence of others. To help identify and prioritize these values, several exercises can be transformative, guiding you toward a clearer understanding of the principles that shape your life.

1. Values Listing Exercise: Start by reviewing a list of common core values, such as **accountability, authenticity, balance, bravery, compassion, creativity, curiosity, dedication, discipline, dignity, empathy, empathy, fairness, fairness, family, honesty, integrity, kindness, patience, respect, self-respect, work-ethics, and trustworthiness**. From this list, select the values that resonate most deeply with you. Try not to overthink or select values you feel you "should" have; instead, focus on what instinctively feels right.

2. Reflective Questions: Asking yourself specific questions can provide insights into your core values. For example:

- When have I felt most fulfilled or proud of myself?
- What achievements or experiences make me feel the most accomplished?
- What qualities do I admire in others, and why?
- When have I felt conflicted or dissatisfied with my choices, and what does that reveal about my values?

3. Narrowing Down: After identifying an initial list of values, prioritize them by considering which ones you absolutely cannot compromise. Narrow your list to three to five core values that feel most essential to who you are. These are the principles that should guide your daily life and major decisions, serving as the foundation for personal and professional growth.

"Values are not just words, values are what we live by. They're about the causes that make our lives worth living." – John Ashcroft

These exercises help you gain clarity on what drives you and serve as a reference point for future decisions. Knowing your core values doesn't mean you have to limit yourself, but it does mean you're setting an intentional foundation for living authentically. Over time, you can revisit and refine these values as you continue to grow and learn.

Chapter: 10
The Seven Pillars of Personal Growth: A Blueprint for Achieving Your Goals

"Craft your blueprint with clarity, build it with determination, and follow it with unwavering persistence. Every step you take brings you closer to the masterpiece of your dreams." - Alex Salem

You must build a strong foundation when embarking on a personal development journey. The following seven pillars will serve as your guiding principles, helping you unlock your full potential and achieve your goals. Each pillar is a stepping stone to a more fulfilling and successful life, fortified by wisdom from great minds who have inspired humanity.

1. Personal Skills

"Personal skills are the keys to unlocking your potential and overcoming life's challenges." – Alex Salem

Personal skills are the bedrock of success, shaping how you interact with the world and navigate its challenges.

"Communication works for those who work at it." - John Powell

- **Communication**: Effective communication is not just about speaking; it's about connection. Whether it's a heartfelt conversation, a persuasive presentation, or a carefully crafted email, your communication ability determines your relationships and opportunities. Authentic communication goes beyond words; it involves active listening, empathy, and understanding. When you connect with others on a deeper level, you build trust and foster meaningful relationships. This skill opens doors to new opportunities, enhances collaboration, and helps you navigate personal and professional landscapes confidently and clearly.

"Lost time is never found again." - Benjamin Franklin

- **Time Management**: Time is your most precious resource, and how you use it defines your success. Prioritize tasks, eliminate distractions, and focus on what truly matters. Mastering time management will empower you to achieve your goals efficiently and effectively. Make every moment count by setting clear priorities, staying disciplined, and dedicating your energy to the activities that bring you closer to your aspirations.

"Leadership is not about being in charge. It is about taking care of those in your charge." – Simon Sinek

- **Leadership**: True leadership is about sparking a fire within others to dream bigger, learn continuously, and achieve greatness. A True leader doesn't merely delegate tasks; they ignite passion, foster innovation, and create a culture of collaboration. They inspire their team to reach new heights, turning visions into reality and challenges into opportunities. By leading with enthusiasm and empathy, they transform ordinary moments into extraordinary achievements.

"In the middle of difficulty lies opportunity." - Albert Einstein

- **Problem-solving**: Every challenge is an opportunity in disguise. Embrace each obstacle with a mindset of creativity, adaptability, and confidence. Cultivate the ability to see beyond the immediate issue and envision innovative solutions. Approach problems as puzzles waiting to be solved, and let your curiosity guide you. Developing strong problem-solving skills transforms setbacks into stepping stones and turns adversity into triumph. Remember, the greatest achievements often arise from the most daunting challenges.

2. Personal Growth

"Personal growth is a journey of relentless self-improvement, where each step forward brings you closer to the best version of yourself." – Alex Salem

Personal growth is about becoming the best version of yourself by embracing change and continually expanding your horizons.

"Anyone who stops learning is old, whether at twenty or eighty. Anyone who keeps learning stays young." - Henry Ford

- **Continuous Learning**: The pursuit of knowledge is the driving force behind both personal and professional growth. It fuels your curiosity, keeps your mind sharp, and opens new horizons. Whether you're diving into books, enrolling in courses, or learning through life experiences, each moment of learning adds to your reservoir of wisdom and skills. Embrace every opportunity to learn; it keeps you innovative and adaptable in an ever-changing world. By continuously seeking knowledge, you enhance your capabilities and inspire those around you to do the same. Remember, the learning journey is endless, and each step forward brings you closer to your fullest potential.

"Becoming is better than being." - Carol Dweck

- **Mindset Development**: Cultivating a growth mindset fundamentally transforms how you approach challenges. Instead of viewing failures as insurmountable obstacles, you begin to see them as valuable stepping stones on your path to success. Embracing a growth mindset means understanding that your abilities and intelligence can be developed through dedication and hard work. This perspective fosters resilience, encourages continuous learning, and inspires you to confidently take on new challenges. By seeing failures as opportunities to learn and grow, you build the courage to push beyond your comfort zone and achieve your full potential. Remember, every setback is a setup for a comeback, and each challenge you face is an opportunity to become stronger and wiser.

"Be like water making its way through cracks... Adjust to the object, and you shall find a way around or through it." - Bruce Lee

- **Adaptability**: The ability to embrace change is crucial for building resilience in our rapidly evolving world. In an era where technological advancements, economic shifts, and societal transformations occur at an unprecedented pace, adaptability becomes a key strength. Embracing change means being open to new ideas, willing to learn, and ready to pivot when necessary. It involves seeing change not as a threat but as an opportunity for growth and innovation. By cultivating this mindset, you equip yourself to handle uncertainties confidently and gracefully. Resilience is developed through flexibility; those who adapt to change are better positioned to thrive in any environment. Remember, change is the only constant in life, and your ability to navigate it determines your success and well-being.

"If your emotional abilities aren't in hand, you won't get very far." Daniel Goleman

- **Emotional Intelligence**: Understanding and managing your emotions while empathizing with others is essential for building deeper connections and making better decisions. Emotional intelligence involves recognizing your feelings, understanding their impact, and regulating them effectively. When you manage your emotions, you respond to situations thoughtfully rather than impulsively. On the other hand, empathy allows you to connect with others on a meaningful level by understanding and sharing their feelings.

This dual capability of self-awareness and empathy fosters trust and strengthens relationships. It also enhances your decision-making process, as you consider not only your own perspective but also the emotions and viewpoints of others. By mastering emotional intelligence, you create a harmonious balance between your inner world and your interactions with those around you, leading to more thoughtful, compassionate, and effective outcomes.

3. Personal Improvement

"Success is not a single act, but a series of consistent actions. What we do repeatedly shapes who we become." - Alex Salem

Continuous improvement builds the discipline and resilience required to achieve greatness.

"A goal is a dream with a deadline." - Napoleon Hill

- **Goal Setting**: Clear goals act as your North Star, providing direction and purpose on your journey to success. When you visualize your objectives, you create a vivid picture of what you want to achieve, making your aspirations more tangible and motivating. Mapping out actionable steps transforms these visions into reality, breaking down your goals into manageable tasks. This process keeps you focused and organized and allows you to track your progress and make necessary adjustments along the way. By setting clear goals and outlining a strategic plan, you navigate your path with confidence and determination, turning your dreams into achievable milestones.

"Discipline equals freedom." - Jocko Willink

- **Self-Discipline**: Mastering your impulses lets you focus on your long-term objectives. Self-discipline is the cornerstone of achieving your goals, enabling you to resist short-term temptations in favor of long-term rewards. By cultivating discipline, you gain control over your actions and decisions, allowing you to stay committed to your plans and make consistent progress. This mastery over your impulses helps you achieve your aspirations and unlocks genuine autonomy, giving you the freedom to shape your life according to your values and ambitions. Remember, discipline is the bridge between goals and accomplishments, guiding you steadily toward success.

"It is health that is real wealth and not pieces of gold and silver." - Mahatma Gandhi

- **Health and Well-Being**: Your physical and mental health forms the bedrock of your ability to thrive in all aspects of life. Prioritizing exercise, nutrition, and mindfulness is essential for maintaining this foundation. Regular physical activity not only strengthens your body but also boosts your mood and energy levels. A balanced diet provides the necessary nutrients to fuel your body and mind, enhancing overall well-being. Mindfulness practices, such as meditation and deep breathing, help you manage stress, stay present, and cultivate a positive mindset. By investing in your health, you equip yourself with the resilience and vitality needed to pursue your goals and enjoy a fulfilling life. Remember, taking care of your body and mind is not just a task but a lifelong commitment to your well-being.

"We all need people who will give us feedback. That's how we improve." - Bill Gates

- **Feedback and Reflection** Regular self-assessment is a powerful catalyst for personal and professional growth. By consistently evaluating your performance and reflecting on your experiences, you gain valuable insights into your strengths and areas for improvement. Seeking constructive feedback from others further enhances this process, providing you with diverse perspectives and actionable advice. Use this feedback to refine your skills and strategies, turning it into a roadmap for continuous improvement. Embrace the mindset that growth is an ongoing journey, and every piece of feedback is an opportunity to learn and evolve. By committing to regular self-assessment and actively seeking feedback, you empower yourself to reach new heights of excellence.

4. Personal Empowerment

"Personal empowerment begins with believing in your own strength and taking charge of your destiny. Embrace your potential and let your inner power shine."- Alex Salem

Personal empowerment is about claiming your inner strength and unleashing your potential.

"No one can make you feel inferior without your consent." - Eleanor Roosevelt

- **Self-Confidence**: Belief in your abilities is the cornerstone of achievement. When you have confidence in your skills and potential, you lay a strong foundation for success. This self-belief fuels your motivation, drives your actions, and helps you overcome obstacles. It empowers you to take risks, embrace challenges, and persist in the face of adversity. By trusting in your capabilities, you unlock the courage to pursue your dreams and turn your aspirations into reality. Remember, every great achievement begins with the unwavering belief that you can accomplish it. Cultivate this belief, and let it guide you on your journey to success.

"Your life changes the moment you make a new, congruent, and committed decision." - Tony Robbins

- **Decision-Making**: Confident decisions stem from a foundation of clarity and thorough preparation. When you take the time to gather information, analyze your options, and understand the potential outcomes, you make informed choices. This clarity allows you to trust your instincts, knowing your decisions are grounded in careful consideration. Taking ownership of your choices means embracing responsibility for the results, whether they lead to success or present new challenges. By preparing diligently and trusting in your abilities, you empower yourself to navigate life's uncertainties confidently and resiliently. Remember, every decision is an opportunity to learn and grow; your confidence will guide you through each step.

"Do not judge me by my success, judge me by how many times I fell down and got back up again." — Nelson Mandela

- **Resilience**: Life is full of challenges and setbacks, but your true strength is revealed in how you respond to them. Each time you face adversity, you have the opportunity to grow stronger and wiser. It's not about avoiding falls but about picking yourself up, learning from the experience, and moving forward with renewed determination and courage. Your resilience is your greatest asset, enabling you to overcome obstacles and achieve your goals despite the difficulties you encounter.

"The question isn't who is going to let me; it's who is going to stop me." - Ayn Rand

- **Autonomy**: Taking control of your life involves more than just making decisions; it means ensuring that your actions are in harmony with your deepest values and long-term goals. This alignment creates a sense of purpose and direction, guiding you through life's challenges and opportunities. When your daily actions reflect what truly matters to you, you build a life that is both fulfilling and meaningful. It requires self-awareness to understand your values, courage to pursue your goals, and discipline to stay on course. You empower yourself to live authentically and achieve your true potential by aligning your actions with your values and goals.

5. Personal Analysis

"Personal analysis is the mirror that reflects your true self, revealing the path to growth and self-improvement." – Alex Salem

Self-awareness is the compass that guides your journey toward personal mastery.

"Self-assessment and attempts at self-improvement are essential aspects of 'the good life.'" – Robert A. Burton

- **Self-Assessment**: Regularly assessing your strengths, weaknesses, and opportunities is crucial for personal and professional growth. By taking the time to reflect on these aspects, you gain a deeper understanding of where you excel and where you need to improve. This self-awareness allows you to leverage your strengths effectively, maximizing your potential and performance. At the same time, it helps you identify areas that require development, enabling you to take proactive steps to address them. Embracing this continuous cycle of evaluation and improvement ensures that you are always moving forward, adapting to new challenges, and seizing growth opportunities. This practice enhances your skills and abilities and builds your confidence and resilience, empowering you to achieve your goals and aspirations.

"When your values are clear to you, making decisions becomes easier." - Roy E. Disney

- **Values Clarification**: Understanding your core principles is essential for living a life that is true to who you are. When you have a clear grasp of your fundamental values and beliefs, you create a strong foundation for making decisions and taking actions that reflect your authentic self. This alignment between your principles and actions brings a sense of integrity and coherence to your life, allowing you to navigate challenges with confidence and clarity. It helps you stay true to your path, even when faced with external pressures or difficult choices. By consistently acting in accordance with your core principles, you build a life that is not only fulfilling but also deeply meaningful, as it resonates with your true identity and purpose.

"You do not rise to the level of your goals; you fall to the level of your systems." - James Clear

- **Behavioral Patterns**: Identifying habits that either support or obstruct your progress is a crucial step in personal development. By taking a close look at your daily routines and behaviors, you can pinpoint which habits are beneficial and which ones are detrimental to your goals. Once you recognize these patterns, you can take deliberate action to replace negative habits with positive ones. This process involves being mindful of your actions, understanding their impact, and making conscious choices to foster habits that contribute to your growth and success. Over time, these positive changes can lead to significant improvements in your overall well-being and productivity, helping you to achieve your aspirations more effectively.

"You don't have to be great to start, but you have to start to be great." - Zig Ziglar

- **Progress Tracking**: Celebrate small victories along the way to maintain your motivation and focus. Recognizing and appreciating these incremental achievements is essential for sustaining your enthusiasm and drive. Each small victory serves as a reminder of your progress, reinforcing your commitment to your goals. By celebrating these moments, you create positive reinforcement that fuels your determination to keep moving forward. This practice not only boosts your morale but also helps you stay focused on the bigger picture, making the journey toward your ultimate objectives more enjoyable and rewarding. Remember, every step forward, no matter how small, brings you closer to your aspirations.

6. Personal Objectives

"Personal objectives are the milestones on your journey to success; clarity and commitment turn them into reality." – Alex Salem

Achieving personal objectives requires a clear vision and a well-thought-out plan.

"If you are working on something exciting that you really care about, you don't have to be pushed. The vision pulls you." – Steve Jobs

- **Vision Creation**: A compelling vision provides your life with purpose and direction. It acts as a guiding star, illuminating the path you need to take to achieve your dreams and aspirations. When you have a clear and inspiring vision, it becomes easier to make decisions and prioritize actions that align with your long-term goals. This sense of purpose fuels your motivation, helping you to stay focused and resilient in the face of challenges. A well-defined vision not only clarifies what you want to achieve but also why it matters, giving your efforts meaning and significance. By continually striving towards your vision, you create a life that is both fulfilling and impactful.

"Plan for what is difficult while it is easy, do what is great while it is small." - Sun Tzu

- **Strategic Planning**: To turn your vision into reality, break it down into manageable, actionable steps. Start by defining your ultimate goal and identifying key milestones. Break each milestone into specific tasks you can tackle systematically. This creates a clear roadmap, making the process less overwhelming and more achievable. Focus on one task at a time to maintain momentum and make consistent progress. Regularly review and adjust your plan to stay on track and adapt to changes. This methodical approach keeps you focused, organized, and motivated, transforming your aspirations into achievable actions and guiding you steadily toward your ultimate goal.

"The key is not to prioritize what's on your schedule, but to schedule your priorities." – Stephen Covey

- **Prioritization**: To make the most of your time and energy, it's crucial to focus on what truly matters. Start by identifying your most important tasks and goals, those that align with your core values and have the most significant impact on your success and well-being. By concentrating your efforts on these high-priority activities, you ensure that your energy is directed toward meaningful and productive outcomes. This process involves making deliberate choices about where to invest your time and resources, often requiring you to say no to less important tasks and distractions. It also means regularly reassessing your priorities to stay aligned with your evolving goals and circumstances. Effective prioritization helps you maintain clarity and focus, reduces stress, and enhances your ability to achieve your objectives efficiently. By dedicating yourself to what truly matters, you create a more purposeful and fulfilling life.

"Celebrate the small wins, and they will eventually add up to something greater." – Unknown

- **Milestone Setting**: Celebrate each milestone as a reminder of your progress. Every small victory you achieve is a testament to your hard work and dedication. These moments of success, no matter how minor they may seem, are crucial in building the momentum needed for more significant achievements. By acknowledging and celebrating these milestones, you reinforce your commitment to your goals and boost your motivation. This positive reinforcement helps you stay focused and energized, making the journey towards your ultimate objectives more enjoyable and rewarding. Remember, each small step forward brings you closer to your larger aspirations, so take the time to appreciate and celebrate your progress along the way.

7. Personal Power

"Personal power is the ability to channel your inner strength and confidence to shape your destiny and inspire those around you." -Alex Salem

Personal power is about harnessing your inner strength and using it to create a positive impact.

"Believe in yourself! Have faith in your abilities!" - Norman Vincent Peale

- **Self-Confidence**: Confidence empowers you to take risks and seize opportunities. When you believe in yourself and your abilities, you are more willing to step out of your comfort zone and embrace new challenges. This self-assurance gives you the courage to pursue your goals, even when the path is uncertain or difficult. Confidence helps you to see opportunities where others might see obstacles, and it motivates you to take decisive action. By trusting in your potential, you can navigate risks with resilience and turn possibilities into achievements. Ultimately, confidence is the key to unlocking your full potential and creating a life filled with growth and success.

"Leadership is influence, nothing more, nothing less."- John Maxwell

- **Influence**: Inspire and uplift others through your actions and words. Your behavior and speech have the power to influence those around you profoundly. By consistently demonstrating kindness, integrity, and positivity, you set an example that others can follow. Your words can offer encouragement, support, and motivation, helping others to see their own potential and believe in their abilities. Similarly, your actions can show what is possible, providing a tangible model of perseverance, resilience, and success. When you inspire and uplift others, you create a ripple effect of positivity and empowerment, fostering a supportive and thriving community. Remember, every kind word and thoughtful action contributes to a better world.

"There's boldness in being assertive; there's strength and confidence." – Bryan Cranston

- **Assertiveness**: Express your thoughts and needs while respecting others. Confident communication fosters mutual respect and understanding. It's important to share your thoughts and needs openly while being considerate of others. When you communicate with confidence and respect, it helps build mutual respect and understanding. This means listening actively, acknowledging different perspectives, and expressing yourself clearly and kindly. By doing so, you create a positive environment where everyone feels valued and heard.

"You have power over your mind – not outside events. Realize this, and you will find strength." - Marcus Aurelius

- **Inner Strength**: Building inner strength involves cultivating resilience by prioritizing your mental and emotional well-being. This means engaging in self-care activities like regular exercise, healthy eating, and adequate sleep to rejuvenate your mind and body. Practicing mindfulness and meditation helps you stay grounded and manage stress while surrounding yourself with supportive and positive people provides essential emotional support. Recognizing and understanding your emotions through journaling or talking to a trusted friend or therapist can offer valuable insights. Embracing a growth mindset allows you to view challenges as opportunities for personal development. By nurturing these aspects of your well-being, you develop the inner strength needed to face life's challenges with resilience and confidence.

"The willingness to accept responsibility for one's own life is the source from which self-respect springs." - Joan Didion

- **Responsibility**: Taking ownership of your choices and their consequences is a powerful act of self-empowerment. When you acknowledge that your actions and decisions directly impact your life, you gain control over your future. This means accepting both the positive outcomes and the challenges that arise from your choices. By doing so, you learn from your experiences, make informed decisions, and grow as an individual. Embracing responsibility allows you to steer your life in the direction you desire, fostering a sense of agency and purpose. Ultimately, it empowers you to shape your destiny and create the life you envision.

"Shoot for the moon. Even if you miss, you'll land among the stars." - Les Brown

- **Motivation**: Harnessing your passion and energy is essential for driving your journey towards success and fulfillment. Passion is the inner fire that motivates you to pursue your goals enthusiastically and determined. When you channel this energy into your endeavors, it propels you forward, even in the face of challenges. To harness your passion effectively, identify what truly excites and inspires you. This could be a career, a hobby, or a cause you deeply care about. Combining your passion with sustained energy creates a powerful force that fuels your journey. This approach helps you achieve your objectives and brings a sense of joy and purpose to your life.

Embracing these seven pillars lays the groundwork for a life filled with purpose, growth, and fulfillment. Remember, the journey of personal development is a marathon, not a sprint. Take one step at a time, stay committed, and trust your ability to transform your life.

Chapter 11: Discovering Your Passions and Interests

The Importance of Passion: Living a Life Filled with Purpose and Excitement

"It is your passion that empowers you to be able to do that thing you were created to do." T. D. Jakes

Passions and interests bring vitality to life, giving us a sense of purpose and excitement that fuels personal growth and fulfillment. Passion is more than just enjoyment; it's a deep connection to activities, ideas, or causes that resonate with our core values and beliefs. When we engage in things that truly excite us, we feel energized, driven, and fully engaged. Discovering these passions is essential for leading a life that feels meaningful and inspiring.

However, finding one's true passion is rarely straightforward. For many, it requires a journey of exploration and experimentation involving curiosity, openness, and a willingness to step out of familiar routines. Rather than expecting passions to reveal themselves immediately, this journey encourages us to try new experiences, connect with different people, and explore various fields. Each new experience brings insights, helping us uncover what truly excites us and gives our life purpose.

Understanding what excites us also enhances our **resilience**. Pursuing passion gives us a sense of purpose, which serves as a source of motivation even in challenging times. When we know what drives us, we become better equipped to handle setbacks, using our enthusiasm to push forward despite obstacles. Passion, therefore, isn't just a source of joy; it's a source of strength and energy, helping us navigate life's ups and downs with focus and determination.

"Passion is the genesis of genius." – Tony Robbins

This chapter will guide readers in discovering their own passions and interests by exploring a wide range of fields, opportunities, and experiences. By embracing curiosity and allowing ourselves to experiment without judgment, we can find activities that align with our values and aspirations, leading us toward a life filled with excitement and purpose.

Exploring Fields and Opportunities: The Power of Curiosity

"The only way to do great work is to love what you do." – Steve Jobs

Discovering our passions often requires stepping outside our comfort zones and exploring diverse fields and opportunities. This exploration is guided by **curiosity**, an essential component of personal growth. Curiosity invites us to ask questions, try new things, and approach the unknown with an open mind. When we're curious, we're less afraid of making mistakes or facing uncertainty; instead, we see each experience as a chance to learn something valuable about ourselves.

One effective way to start is by **sampling various activities and fields**. This could mean taking up a new hobby, volunteering for a cause, enrolling in an online course, or connecting with professionals in different industries. Even seemingly unrelated activities can spark inspiration and reveal unexpected interests. For example, someone interested in writing might find that they enjoy photography, and together, these interests might lead to a passion for storytelling through visual and written media.

Curiosity also encourages us to view **failure as a learning opportunity**. Exploring new fields doesn't always lead to immediate passion; sometimes, it simply shows us what we don't enjoy. However, each experience—whether it's a success or a setback—helps refine our understanding of what resonates with us. By treating exploration as an ongoing journey rather than a checklist, we keep our minds open to the possibilities that lie ahead.

"We keep moving forward, opening new doors, and doing new things, because we're curious and

curiosity keeps leading us down new paths." – Walt Disney

As we explore, it's important to be mindful of the activities that make us lose track of time or feel particularly fulfilled. These moments often hint at passions that, when nurtured, can evolve into meaningful pursuits. By remaining curious and adventurous, we create a life that is rich with experiences and insights, bringing us closer to understanding what truly excites and motivates us.

Embracing a Wide Range of Experiences: Stepping Beyond Comfort Zones

"Life begins at the end of your comfort zone." – Neale Donald Walsch

To discover our true passions, it's essential to embrace a range of experiences, even those that may initially feel uncomfortable or unfamiliar. Growth happens when we step outside our comfort zones, challenging ourselves to see life from new perspectives and encounter situations that stretch our capabilities. Often, the activities that we think we might dislike or feel apprehensive about can end up sparking deep interest and excitement once we give them a try.

"Make a conscious effort to embrace new experiences, for they ignite creativity and open doors to endless possibilities." - Alex Salem

One way to embrace new experiences is to make a conscious effort to say **"yes" more often**—especially to opportunities that encourage learning and creativity. For instance, joining a public speaking group, taking an art class, learning a new language, learning to sew, playing an instrument, crafting various items, or trying out a new sport can all lead to unexpected revelations. Each new experience serves as a way to broaden our skills and knowledge, which, in turn, opens doors to passions we may not have considered.

Another important aspect of exploring is **connecting with people who have diverse passions**. Learning from others can introduce us to fields and interests we hadn't thought to explore. Conversations with friends, mentors, or community members can inspire new ideas, reveal hidden opportunities, and provide support as we venture into unfamiliar territory. These connections help us see the world through others' eyes, often sparking our own curiosity and leading us to passions we may not have otherwise discovered.

"Don't be afraid to give up the good to go for the great." – John D. Rockefeller

When we embrace a wide range of experiences, we develop a well-rounded perspective on life and a deeper understanding of ourselves. Some activities may become lifelong interests, while others may simply serve as lessons along the way. Regardless, each experience contributes to our personal growth, helping us refine our passions and understand what truly brings us joy.

Overcoming Obstacles to Pursuing Passions: Facing Fear and Self-Doubt

"The greatest barrier to success is the fear of failure." – Sven Goran Eriksson

Even once we identify our passions, pursuing them can be challenging. **Fear of failure**, self-doubt, and societal expectations often stand as obstacles that prevent us from fully embracing what excites us. These hurdles are natural, as venturing into something we care deeply about can feel risky. However, recognizing and addressing these obstacles is essential for making progress on our journey to a fulfilling life.

"Fear of judgment can hold us back, but your journey is uniquely yours. Courage to follow your own path defines you." - Alex Salem

One common challenge is the **fear of judgment**—the worry that others may not understand or appreciate our choices. This fear can make us hesitate, especially if our passion deviates from traditional or socially accepted paths. To overcome this, it's helpful to remind ourselves that passion is deeply personal; it doesn't require validation from others. When we focus on what excites us and brings joy, we create an inner motivation that withstands external judgments.

"The only limit to our realization of tomorrow will be our doubts of today." - Franklin D. Roosevelt

Another obstacle is the feeling of **self-doubt**. Believe in your beliefs and doubt your doubts. Don't believe in your doubts and doubt your beliefs. When faced with the unknown, it's easy to question our abilities or think, "I'm not good enough to succeed at this." However, pursuing a passion is not about perfection; it's about commitment, growth, and learning. Starting with small, manageable steps can build confidence over time, allowing us to progress without becoming overwhelmed by unrealistic expectations.

"Success by the yard is hard, but success by the inch is a cinch" - John Bytheway

Time constraints and practical limitations also present challenges. Many people feel that their schedules are too full to dedicate time to new interests. But by prioritizing what truly matters, we can often make room for our passions, even if only in small increments. Whether it's dedicating 10 minutes a day to writing, setting aside an hour a week for a new hobby, or planning a weekend workshop, small, consistent efforts add up and can lead to substantial progress.

Ultimately, the willingness to face and overcome these obstacles strengthens our resolve and deepens our connection to our passions. By confronting fear and doubt, we become more resilient, ensuring that our interests and pursuits continue to grow with us over time.

Chapter 12: Cultivating a Growth Mindset

Understanding the Growth Mindset: Embracing Change and Possibility

"Whether you think you can or think you can't, you're right." – Henry Ford

A **growth mindset** is the belief that abilities and intelligence can be developed through dedication, learning, and effort. In contrast, a **fixed mindset** holds that abilities are static and cannot be changed. This fundamental difference shapes how we approach challenges, handle setbacks, and pursue personal development. Adopting a growth mindset transforms the way we see ourselves and our potential, allowing us to embrace challenges as opportunities and to learn from experiences, no matter the outcome.

"Every challenge or obstacle is a steppingstone, guiding you towards greater strength and resilience." – Alex Salem

With a growth mindset, we shift our focus from simply proving our abilities to expanding them. This outlook fosters resilience, as we see obstacles not as barriers but as steps on the journey to improvement. For example, someone with a fixed mindset might view a difficult project as a test of their inherent skill, fearing failure and perhaps avoiding the challenge. However, a person with a growth mindset would see the same project as a chance to learn new skills, gain experience, and become more capable through the process.

"Success is not an accident; it's a choice." – Stephen Curry

By adopting a growth mindset, we open ourselves to a world of possibility where self-improvement is continuous. This mindset empowers us to take control of our progress, no longer limited by the belief that we are confined to our current skills or knowledge. In this chapter, readers will explore the practical steps to develop a growth mindset, understand the benefits it offers, and learn how to shift from a fixed mindset to one that embraces growth and transformation.

Growth Mindset vs. Fixed Mindset: Key Differences and Why They Matter

"The mind is everything. What you think, you become." – Buddha

The distinction between a growth mindset and a fixed mindset affects nearly every area of our lives, from our approach to learning and relationships to how we handle challenges and setbacks. Understanding these differences provides clarity on how adopting a growth mindset can empower us to reach our potential.

"A fixed mindset limits your potential; a growth mindset expands it." - Anonymous

In a **fixed mindset**, individuals believe their abilities, intelligence, and talents are unchangeable. They often view failure as a reflection of their inherent limitations, making them more likely to avoid challenges, give up easily, or feel threatened by others' success. This mindset fosters a sense of inadequacy and can stifle growth, as individuals may feel that their potential is predetermined.

Conversely, a **growth mindset** sees failure as an essential part of learning and views effort as a pathway to mastery. People with this mindset understand that abilities are developed through hard work, perseverance, and learning from mistakes. Rather than avoiding difficult tasks, they welcome challenges, seeing them as opportunities to grow. A growth mindset creates a sense of resilience, helping individuals bounce back from setbacks and keep striving toward improvement.

"Failure is simply the opportunity to begin again, this time more intelligently." – Henry Ford

By recognizing these differences, we can start shifting our mindset to embrace growth, realizing that setbacks are not failures but stepping stones. This shift allows us to take on new challenges, build resilience, and develop the confidence to pursue our goals. Developing a growth mindset encourages a lifelong journey of self-improvement and equips us with the determination to keep moving forward, even when faced with obstacles.

The Benefits of a Growth Mindset: Determination, Resilience, and Success

"Believe you can, and you're halfway there." – Theodore Roosevelt

Adopting a growth mindset brings numerous benefits, particularly in the areas of resilience, determination, and overall success. With a growth mindset, we not only become more willing to face challenges but also develop the mental stamina needed to overcome them. This outlook allows us to stay focused on our goals, viewing setbacks as opportunities to strengthen our skills rather than as indicators of failure.

One of the key benefits of a growth mindset is **increased resilience**. When we believe that our abilities can grow through effort, we become better equipped to handle difficulties and remain persistent in the face of adversity. This resilience is crucial for personal growth, as it enables us to continue pursuing our passions and goals even when circumstances are challenging.

A growth mindset also fosters **determination**. When we view effort as an essential part of learning, we are more likely to embrace hard work and dedication, staying committed to our objectives even when progress seems slow. Determination is fueled by the understanding that growth happens gradually, and each step forward brings us closer to our potential.

"Our greatest glory is not in never falling, but in rising every time we fall." – Confucius

Furthermore, a growth mindset creates a sense of **ownership over personal success**. By acknowledging that we have the power to improve and learn, we build confidence and take pride in our accomplishments, knowing they result from our hard work and perseverance. This sense of ownership enhances our self-esteem and motivates us to keep pushing forward, continually reaching for new levels of achievement.

By embracing a growth mindset, we cultivate an approach to life that is proactive, determined, and resilient—qualities that support us on our journey toward fulfilling our potential.

Practical Exercises to Foster a Growth Mindset

"You don't have to be great to start, but you have to start to be great." – Zig Ziglar

Cultivating a growth mindset requires more than just an intellectual shift; it involves practical steps and consistent efforts to reshape our habits and attitudes. Here are some exercises that can help foster a growth-oriented approach, enabling us to embrace challenges, learn from feedback, and maintain focus on our personal growth.

"It is a tragedy to journey through life without challenging the limits of our mind, soul, and body to discover their true potential." – Alex Salem

1. Embrace Challenges: Start by consciously seeking out tasks that push you beyond your comfort zone. These might include learning a new skill, taking on a project that feels ambitious, or trying something you've always been hesitant about. Each challenge is an opportunity to test your limits and see growth as a result. Reflect on the experience afterward, noting any difficulties you encountered and what you learned in the process.

2. Practice Self-Reflection: Reflection is essential to developing a growth mindset. Each day, take a few minutes to think about any obstacles you faced, what you did well, and areas where you could improve. Writing down your reflections can be especially useful, as it allows you to track your progress over time and identify patterns in your thoughts and actions. This exercise reinforces the idea that improvement is an ongoing journey.

3. Reframe Failures as Learning Opportunities: Whenever you encounter setbacks, try viewing them as valuable lessons rather than reasons to give up. Ask yourself: What did this experience teach me? How can I apply this knowledge to future situations? By reframing failure in a positive light, you reinforce the belief that mistakes are part of the growth process.

"Every failure is a stepping stone to success." – Arianna Huffington

4. Value Constructive Criticism: Instead of taking feedback personally, try to view it as an opportunity to grow. Constructive criticism offers insights into areas for improvement that may not be visible to you. When receiving feedback, focus on how you can apply it to develop further, understanding that criticism is a powerful tool for enhancing your skills and mindset.

Each of these exercises encourages a shift in perspective, helping to replace self-doubt with determination and resilience. By practicing these strategies regularly, you can strengthen your growth mindset and keep moving toward your personal and professional goals with renewed confidence.

Using a Growth Mindset to Guide Goal Achievement

"You are never too old to set another goal or to dream a new dream." – C.S. Lewis

A growth mindset not only enhances our outlook but also transforms how we approach and achieve our goals. When we view our objectives through a growth-oriented lens, we become more resilient, adaptable, and focused on long-term progress. This mindset shifts our goals from being endpoints to becoming stepping stones in our journey of continuous improvement.

"The man who moves a mountain begins by carrying away small stones." – Confucius

To effectively pursue goals with a growth mindset, it's important to **break down larger objectives into manageable steps**. By creating smaller, achievable milestones, we can track our progress and celebrate incremental successes, reinforcing our motivation and reminding us of how far we've come. Each small achievement builds momentum, making the overall goal feel attainable and within reach.

"Set goals not for the outcome itself, but for who you get to become in the process." Jim Rohn

Another powerful approach is to **focus on the process, not just the outcome**. When we set our sights solely on the end goal, we may become discouraged by obstacles along the way. However, if we concentrate on learning, improving, and growing throughout the process, each step becomes valuable regardless of the final result. This focus on the journey rather than the destination ensures that we remain engaged, curious, and resilient, even when challenges arise.

"Self-compassion allows us to embrace our imperfections and see them as part of the shared human experience, rather than personal defects." – Kristin Neff

Lastly, it's important to **practice self-compassion** in goal achievement. A growth mindset doesn't mean pushing ourselves relentlessly; it means understanding that setbacks are part of progress and treating ourselves kindly when we fall short. Self-compassion allows us to recover quickly from mistakes, learn from them, and continue moving forward without self-criticism holding us back.

By cultivating a growth mindset, we transform goal-setting into a meaningful process of personal development. Each goal achieved is not just an accomplishment; it's a testament to our commitment to growth, resilience, and self-discovery.

Chapter 13: Setting Personal Growth Goals

The Power of Goal Setting: Turning Dreams into Actionable Plans

"Dreams written down with specific dates become goals, goals broken down into steps become plans, plans backed by actions turns your dream into reality" – Alex Salem

Personal growth goals are the stepping stones that lead us toward a life aligned with our values, dreams, and potential. While dreams provide inspiration, goals turn that inspiration into action. Setting personal growth goals empowers us to transform abstract aspirations into tangible milestones, giving us a clear direction and a sense of purpose. When we set goals, we're not only identifying what we want but actively committing to the process of becoming who we want to be.

The significance of goal setting lies in its ability to give structure to our ambitions. Goals create focus and motivation, helping us prioritize our time, energy, and resources on the things that truly matter. Without clear goals, it's easy to drift through life, reacting to circumstances rather than taking proactive steps. Goals keep us grounded, providing measurable progress markers that encourage us to keep moving forward, even when challenges arise.

Personal growth goals differ from typical objectives because they are centered on self-improvement rather than external achievements. While professional goals may focus on career advancement or financial success, personal growth goals are often about enhancing qualities like resilience, empathy, knowledge, or emotional well-being. Setting these goals is a declaration that we are dedicated to evolving, not just achieving.

"The journey of a thousand miles begins with one step." – Lao Tzu

This chapter will explore effective strategies for setting meaningful, achievable personal growth goals. By understanding how to break down goals into manageable steps and incorporate them into our daily routines, we can make personal growth a continuous and fulfilling journey.

Tips for Setting Realistic and Meaningful Personal Growth Goals

"The future depends on what you do today." – Mahatma Gandhi

Setting realistic goals is key to maintaining motivation and ensuring steady progress on the path to personal growth. Goals that are too ambitious can lead to burnout or discouragement, while goals that are too easy may not push us toward meaningful change. Striking a balance requires a thoughtful approach to goal setting, ensuring that each objective is both attainable and aligned with our true aspirations.

One effective strategy is to use the **SMART criteria**—Specific, Measurable, Achievable, Relevant, and Time-bound. SMART goals provide a framework that turns vague intentions into clear plans. For example, instead of setting a broad goal like "improve my communication skills," a SMART goal would be "attend a public speaking workshop and practice presenting once a week for three months." This level of specificity makes the goal tangible and gives us a roadmap for success.

Another tip is to **align goals with your core values**. Personal growth is most fulfilling when it resonates with who we are at our core. Before setting a goal, consider how it fits into your broader values and life vision. If growth in creativity is a core value, for instance, setting a goal to learn a new artistic skill might feel more satisfying and meaningful than unrelated goals. Goals rooted in our values motivate us on a deeper level, making it easier to stay committed even when the journey gets tough.

"Define your goals and watch your dreams become reality. Start today and embrace the extraordinary journey ahead!" – Alex Salem

Finally, it's essential to **start small**. Breaking larger aspirations into smaller, manageable steps creates a clear path forward and allows us to celebrate incremental successes. For example, if your ultimate goal is to develop emotional resilience, break it down into specific steps like "practice mindfulness for 10 minutes daily," "read a book on emotional intelligence," and "journal about challenges once a week." Small steps make big goals achievable, helping us maintain momentum and see progress as we move closer to our aspirations.

Breaking Goals into Actionable Steps

"A goal is not always meant to be reached; it often serves simply as something to aim at." – Bruce Lee

One of the most effective ways to make progress on personal growth goals is to break them down into smaller, actionable steps. This process transforms large, daunting goals into manageable tasks that can be worked on gradually. By tackling one small step at a time, we maintain a steady pace and prevent feelings of overwhelm, making it more likely that we'll stick to our plan and achieve long-term success.

To break down a goal, start by **identifying the end result** and then working backward. For example, if the goal is to "develop a regular exercise routine," think about what that routine would look like in its final form—perhaps working out three times a week. Next, outline the steps needed to get there, such as researching types of exercise, scheduling time each week, and gradually increasing workout frequency. This approach gives a clear path, making each step a natural progression toward the goal.

Prioritization is also key when breaking down goals. Not all steps will be equally important or require the same amount of effort. To stay organized, rank each step in order of priority and start with the most critical actions. This allows for focused effort, helping us build momentum on the steps that make the greatest impact.

"If you cannot do great things, do small things in a great way." – Napoleon Hill

Creating a **timeline** for each step further enhances accountability. Setting a date for when each action should be completed helps prevent procrastination and keeps us on track. By setting realistic timelines and focusing on one small step at a time, we establish a rhythm that encourages continuous progress, making even the largest goals feel attainable and within reach.

Incorporating Personal Growth Strategies into Daily Life

"Success is the sum of small efforts, repeated day in and day out." – Robert Collier

To make lasting progress toward personal growth goals, it's essential to integrate growth strategies into our daily routines. Consistency is key to transformation, and by incorporating personal development practices into our everyday life, we build habits that support continuous improvement. This approach not only makes goals more achievable but also turns personal growth into a natural, ongoing part of life.

"We first make our habits, and then our habits make us" - John Dryden

Daily habits are a powerful way to maintain focus on our goals. For example, if one of your personal growth goals is to cultivate mindfulness, establish a simple daily habit like practicing meditation each morning or taking five minutes at the end of the day to reflect on positive experiences. Small, consistent actions like these accumulate over time, making significant growth feel effortless.

Another effective strategy is to **create visual reminders**. Keeping your goals visible reinforces your commitment to them and acts as a motivator. Write down your goals and place them somewhere you'll see often, such as on your desk or in a journal. You could also use digital tools like phone reminders or habit-tracking apps to keep your focus sharp. Seeing progress in real-time can be a huge motivator, reminding you of how far you've come and encouraging you to stay the course.

"The secret of getting ahead is getting started." – Mark Twain

Accountability partners can also be valuable for goal-setting. Sharing your goals with a friend, mentor, or coach creates a layer of accountability, making it more likely that you'll follow through. Regular check-ins with an accountability partner provide opportunities to discuss challenges, celebrate successes, and adjust your strategy as needed. This support network not only keeps you motivated but also brings a sense of camaraderie to your growth journey.

Incorporating personal growth strategies into daily life creates a steady momentum, transforming big goals into achievable actions and making self-improvement an integral part of every day.

PART III:
Achieving Balance and Wellness in a Technology-Driven World

Chapter 14: Physical Well-Being

The Foundation of Physical Well-Being: Fueling Your Journey

"Take care of your body. It's the only place you have to live." – Jim Rohn

Physical well-being is the cornerstone of a fulfilled and productive life. When we prioritize our physical health, we not only increase our energy levels and resilience but also empower ourselves to pursue our goals with greater focus and determination. Physical well-being is about more than fitness or appearance; it's about nurturing our bodies in ways that support mental clarity, emotional stability, and long-term health.

Achieving physical well-being begins with understanding that our bodies are incredibly intuitive and responsive. They communicate their needs through signals like hunger, fatigue, or tension, and learning to listen to these cues is essential for maintaining balance. By tuning in to our bodies, we can make conscious choices that promote vitality and well-being rather than simply following routines out of habit.

One of the most fundamental aspects of physical well-being is **nutrition**. A balanced diet provides the fuel necessary for both body and mind to function optimally. Nutrition affects not only our physical health but also our mood, energy levels, and cognitive abilities. Rather than viewing food as a source of calories alone, consider it as a source of energy and nourishment that enables us to reach our full potential.

"Let food be thy medicine and medicine be thy food." – Hippocrates

This chapter will explore key areas of physical well-being, including balanced nutrition, regular exercise, and the importance of rest and recovery. By adopting a holistic approach to physical health, readers can cultivate a lifestyle that supports both their personal growth and long-term well-being.

Listening to Your Body: Recognizing and Responding to Its Needs

"Your body hears everything your mind says." – Naomi Judd

A key aspect of physical well-being is learning to listen to our bodies. Our bodies constantly send us signals—whether it's hunger, fatigue, or tension—telling us what they need to maintain balance and function optimally. Developing an awareness of these cues allows us to make mindful choices that support our health and energy, preventing burnout and enhancing our overall quality of life.

To cultivate this awareness, start by paying attention to **how different foods, activities, and routines make you feel**. For instance, notice the effect that a healthy breakfast has on your energy and focus throughout the morning. Similarly, observe how certain types of exercise or relaxation techniques influence your mood and productivity. This self-awareness empowers you to make choices that enhance your well-being based on your unique needs.

Another critical component of listening to your body is understanding when it needs **rest and recovery**. It's common to push ourselves to the point of exhaustion in pursuit of our goals, but this approach is unsustainable. Just as physical training requires periods of rest for muscles to rebuild and strengthen, our bodies and minds need breaks to recharge. By honoring our need for rest, we allow ourselves to maintain a steady pace without risking burnout.

"Rest and self-care are so important. When you take time to replenish your spirit, it allows you to serve others from the overflow. You cannot serve from an empty vessel." – Eleanor Brown

Listening to your body is an essential practice that supports physical health and overall well-being. It encourages us to make adjustments based on real needs rather than rigid routines, creating a balanced lifestyle that supports sustained energy, resilience, and growth.

The Power of Nutrition: Fueling Your Body and Mind

"To eat is a necessity, but to eat intelligently is an art." – François de La Rochefoucauld

Nutrition is the foundation of physical well-being. The foods we consume have a profound impact on our energy, mental clarity, mood, and overall health. Eating a balanced diet is about providing our bodies with the nutrients they need to perform at their best, rather than simply counting calories or following fad diets. When we fuel our bodies well, we enable them to support us in pursuing our goals and maintaining resilience in the face of daily challenges.

"If nature doesn't make it, don't take it." - Anonymous

A balanced diet includes a variety of **whole foods**—fruits, vegetables, lean proteins, whole grains, and healthy fats. Each food group plays a unique role in supporting bodily functions, from repairing muscle tissue to strengthening the immune system. Carbohydrates, for instance, are a primary source of energy, while proteins support muscle repair and growth. Healthy fats are crucial for brain function, while vitamins and minerals help regulate bodily processes.

In addition to focusing on nutrient-dense foods, hydration is an essential part of nutrition. **Water is vital for nearly every bodily function**, from digestion and circulation to temperature regulation and joint health. Staying hydrated helps maintain focus, reduces fatigue, and promotes physical endurance. A simple way to ensure adequate hydration is to carry a water bottle and drink consistently throughout the day, especially during and after physical activity.

"You are what you eat, so don't be fast, cheap, easy, or fake." – Unknown

Cultivating healthy eating habits can transform not only your physical health but also your mental well-being. By choosing foods that nourish rather than deplete, you provide your body and mind with the foundation needed to achieve your goals and maintain a balanced lifestyle.

Here are some healthy eating habits you can incorporate into your daily routine to support your overall well-being and make them lasting habits:

Cut or Limit Sugary Drinks: Cut down on beverages with added sugars and choose water, unsweetened coffee or tea, and even 100% fruit juice in moderation.

Choose Whole Grains: OPT for whole wheat bread, brown rice, wild rice, forbidden rice, quinoa, teff, and oats. Look for foods labeled "whole grain" or "whole wheat" in the ingredients list.

Snack on Fruit: Keep fresh, frozen, or dried fruits handy for nutritious snacks throughout the day.

Eat More Veggies: Aim for a variety of colorful vegetables. Fresh or frozen.

Practice Mindful Eating: Pay attention to your hunger cues and savor each bite. Avoid mindless eating.

Plan Your Meals: Prepare balanced meals ahead of time to avoid unhealthy last-minute choices.

Add Variety to Your Diet: Explore different healthy foods and flavors. Variety ensures you get a wide range of nutrients.

Cut or Limit Processed Meats and Meats: Reduce consumption of processed meats. Choose lean proteins like eggs, beans, turkey, chicken, and fish.

Cut or Limit Fried Foods: Eliminate or Cut back on fried foods, which are often high in unhealthy fats.

Stop or Drink Less Alcohol: Moderation is key. Limit alcohol intake to support overall health.

Eat More Beans: Incorporate legumes like lentils, chickpeas, and black beans. They're rich in fiber and protein.

Eat Breakfast: Start your day with a balanced meal to kick-start your metabolism.

Cut or Decrease Salt Intake: Reduce sodium by using herbs and spices for flavor instead of excessive salt.

Include Healthy Fats: Choose sources like avocados, nuts, seeds, and olive oil.

Eat Your Favorite Foods: Enjoy treats in moderation. Deprivation isn't sustainable.

Emphasize Fruits and Vegetables: Fill half your plate with colorful produce for essential vitamins and minerals.

Give your body a break: It is normal to be hungry; fast once a week or once a month.

Remember, small changes over time can lead to significant improvements in your health. Choose a few habits to start with and gradually build upon them.

The Importance of Exercise: Building Strength, Stamina, and Resilience

"Movement is a medicine for creating change in a person's physical, emotional, and mental states." – Carol Welch

Exercise is a cornerstone of physical well-being, offering benefits that go beyond physical fitness. Regular physical activity strengthens the body, increases stamina, and improves mental resilience. Exercise has been shown to reduce stress, elevate mood, and enhance cognitive function, making it a powerful tool for supporting both physical and mental health. When we exercise, we release endorphins, often referred to as "feel-good" hormones, which boost our sense of well-being and relieve tension.

The type of exercise that's best for you depends on your individual goals, preferences, and current fitness level. **Aerobic activities** like walking, jogging, cycling, or swimming improve cardiovascular health, increasing stamina and endurance. **Strength training** helps build and maintain muscle, supporting bone health and improving balance and coordination. **Flexibility exercises**, like yoga or stretching, enhance range of motion and reduce the risk of injury.

A balanced fitness routine incorporates a mix of these activities tailored to your body's needs. Even modest amounts of physical activity can make a significant difference. For instance, taking a 15-minute walk daily, using stairs instead of elevators, or stretching periodically throughout the day can all contribute to improved physical health. Regular exercise also boosts self-discipline and focus, which can spill over into other areas of personal growth.

Incorporating exercise into daily life doesn't require intense training; it's about consistency and commitment to moving your body in ways that feel good. When we make physical activity a habit, we strengthen our ability to pursue our goals with energy and determination, enhancing both our physical resilience and our mental strength.

Establishing a daily workout and stretching routine can significantly improve your overall health and well-being. Here's a combination of exercises and stretches that you can easily incorporate into your daily routine, no matter where you are:

Cardiovascular Exercise (Cardio): Aim for at least 15 minutes of moderate-intensity cardio most days of the week. Options include brisk walking, jogging, cycling, dancing, or swimming.

Strength Training: Include resistance exercises to build muscle strength. Focus on major muscle groups: legs, arms, chest, back, and core. Use bodyweight exercises (like squats, lunges, push-ups) or weights (dumbbells, resistance bands).

Balance and Stability: Practice balance exercises to improve stability and prevent falls. Examples: single-leg stands, heel-to-toe walking, or yoga poses.

Chapter 15: Mental Well-Being

The Importance of Mental Well-Being: Thriving in a High-Pressure World

"Mental health... is not a destination, but a process. It's about how you drive, not where you're going." – Noam Shpancer

In today's fast-paced, high-pressure world, maintaining mental well-being has become more critical than ever. The demands of work, relationships, and daily responsibilities can leave us feeling overwhelmed, stressed, and disconnected. Mental well-being is not just about the absence of mental illness; it's about cultivating a state of emotional balance, resilience, and clarity that allows us to thrive despite challenges.

Mental well-being is deeply interconnected with our physical health, productivity, and relationships. When we prioritize our mental health, we create a foundation for personal growth, decision-making, and fulfilling connections with others. A healthy mind allows us to approach problems with creativity, stay focused on our goals, and navigate difficulties with confidence.

However, mental well-being requires conscious effort. It involves recognizing the impact of stress on our lives and adopting strategies that help us cope with and manage this stress effectively. Ignoring mental health concerns or neglecting self-care can lead to burnout, anxiety, and other challenges that diminish our ability to function at our best.

"You don't have to control your thoughts. You just have to stop letting them control you." – Dan Millman

This chapter explores the importance of mental well-being and provides actionable strategies to cultivate stability and resilience. From mindfulness practices like meditation to stress management techniques and work-life balance tips, readers will learn how to nurture their mental health and create a life of calm and purpose.

Mindfulness Practices: Cultivating Presence and Peace

"The present moment is filled with joy and happiness. If you are attentive, you will see it." – Thich Nhat Hanh

Mindfulness is one of the most powerful tools for enhancing mental well-being. It is the practice of focusing on the present moment without judgment, allowing us to step away from the chaos of daily life and reconnect with our inner selves. Mindfulness fosters a sense of calm and clarity, helping us manage stress, reduce anxiety, and improve emotional balance.

Meditation is a cornerstone of mindfulness. By dedicating just a few minutes each day to sitting quietly and focusing on the breath, we can cultivate a sense of inner peace and resilience. Meditation trains the mind to let go of distracting thoughts, replacing them with a deep awareness of the present. Over time, this practice helps us respond to stress with greater patience and composure.

Mindfulness doesn't require lengthy sessions or specialized tools; it can be integrated into everyday activities. For example:

- During meals, take time to savor each bite, paying attention to the flavors, textures, and aromas.
- While walking, focus on the sensations of your steps, the feel of the air, and the sights and sounds around you.
- When feeling overwhelmed, pause for a minute to take a few deep breaths, grounding yourself in the present moment.

"You should sit in meditation for 20 minutes a day, unless you're too busy. Then you should sit for an hour." – Zen Proverb

Mindfulness is not about eliminating stress but about changing our relationship with it. By practicing mindfulness, we train ourselves to observe thoughts and emotions without being controlled by them, creating space for calm and intentional action even in the face of challenges.

Stress Management: Turning Pressure into Productivity

"It's not stress that kills us, it is our reaction to it." – Hans Selye

Stress is an inevitable part of life, but how we respond to it determines whether it drains us or drives us. Managing stress effectively involves recognizing its sources and adopting strategies that help us cope without becoming overwhelmed. With the right tools, we can turn stress from a hindrance into an opportunity for growth and productivity.

1. Identify Stressors: The first step in managing stress is to understand its root causes. Are work deadlines piling up? Is your personal life feeling neglected? Journaling about daily stressors can help identify recurring patterns and areas that need attention. Once you know the sources, you can address them systematically rather than feeling consumed by them.

2. Prioritize and Delegate: Often, stress arises from taking on too much at once. Learning to prioritize tasks and delegate responsibilities where possible can reduce the load. Ask yourself, "What is most important right now?" Focus your energy on high-priority tasks while letting go of less critical ones.

3. Practice Physical Relaxation Techniques: Activities like yoga, deep breathing, or progressive muscle relaxation help release physical tension caused by stress. Even taking a short break to stretch or go for a walk can have an immediate calming effect.

"Almost everything will work again if you unplug it for a few minutes, including you." – Anne Lamott

4. Set Boundaries: Establishing boundaries between work and personal life is crucial for maintaining mental well-being. Avoid checking emails late at night or working during personal downtime. Protecting your time for rest and relationships ensures that stress doesn't dominate your life.

Stress is not inherently bad; it can push us to meet challenges and grow. However, unmanaged stress can take a toll on our mental and physical health. By adopting effective stress management techniques, we regain control and create space for balance and growth.

Balancing Work and Personal Life: Creating Harmony Amid Demands

"Never get so busy making a living that you forget to make a life." – Dolly Parton

Striking a balance between work and personal life is essential for maintaining mental well-being. In a world that often glorifies busyness and productivity, it's easy to neglect personal needs, relationships, and self-care in favor of professional demands. However, long-term mental health requires harmony between the two, allowing us to thrive both at work and in our personal lives.

1. Set Clear Boundaries: Defining boundaries between work and personal time is critical. For instance, avoid answering work emails after a designated time in the evening or during weekends. Communicating these boundaries to colleagues and loved ones ensures that your time for relaxation and relationships is respected.

2. Schedule Personal Time: Treat personal time with the same importance as work commitments. Schedule time for activities that bring joy, whether it's pursuing hobbies, spending time with family, or simply relaxing with a good book. This intentional time away from work fosters renewal and reduces the risk of burnout.

3. Learn to Say No: Overcommitment is a common cause of stress. Learning to say no to tasks or responsibilities that do not align with your priorities can free up valuable time for personal growth and rest. It's not about avoiding responsibility but ensuring you're dedicating energy to what truly matters.

"Balance is not something you find; it's something you create." Jana Kingsford

4. Practice the Art of Presence: When at work, focus fully on your tasks; when with loved ones, be fully present. This separation enhances both productivity and personal connection, ensuring that each area of life gets the attention it deserves.

Balancing work and personal life is not about achieving perfection but about making intentional choices that honor your well-being. By creating this balance, you build a sustainable lifestyle that supports both your professional success and personal happiness.

Feed the mind when the body is tired, train the body when the mind is tired; that way, you never get tired.' – Alex Salem

Chapter 16: Nurture Your Soul

The Essence of Spiritual Well-Being: Connecting with Your Inner Self

"The soul always knows what to do to heal itself. The challenge is to silence the mind." – Caroline Myss

Nurturing the soul is about creating a deeper connection with your inner self, fostering peace, and seeking purpose beyond the tangible aspects of daily life. In a world filled with constant demands and distractions, prioritizing spiritual well-being allows us to find clarity, calm, and a sense of meaning. It invites us to explore who we are at our core and align our actions with our truest values.

Spiritual well-being is not tied to any specific religion or belief system; it is a deeply personal journey of self-discovery and reflection. For some, it may involve faith or meditation, while for others, it might be about connecting with nature, engaging in creative expression, or practicing gratitude. The goal is to cultivate a sense of harmony within yourself and the world around you.

Engaging in **self-reflection** is one of the most powerful ways to nurture the soul. Reflecting on your thoughts, feelings, and experiences helps uncover patterns and truths about your inner world. It creates space to understand your motivations, fears, and aspirations, guiding you toward greater self-awareness and fulfillment.

"Through self-reflection, we nurture our souls, finding clarity and peace within ourselves." – Alex Salem

This chapter explores the importance of nurturing the soul and offers practices to help you connect with your inner self. By integrating these practices into daily life, you can cultivate a sense of spiritual well-being that grounds and enriches your life.

Practices for Self-Reflection: Exploring Your Inner World

"Mastering others is strength; mastering yourself is true power." – Lao Tzu

Self-reflection is a cornerstone of nurturing the soul. It allows us to pause, step back from the noise of daily life, and connect with our inner thoughts and feelings. Through reflection, we gain insight into who we are, what drives us, and how we can align our actions with our authentic selves. This practice not only deepens our self-awareness but also fosters clarity and purpose in our lives.

1. Journaling: Writing down your thoughts, emotions, and experiences can be a transformative self-reflection tool. Journaling provides a safe space to explore your inner world without judgment. You can start with prompts like:

- What brings me the most joy in my life right now?
- What challenges am I facing, and what are they teaching me?

- What do I want to let go of, and what do I want to embrace?

2. Quiet Contemplation: Setting aside time for quiet reflection allows you to process your thoughts and connect with your inner self. Find a peaceful environment, close your eyes, and simply let your mind wander. Pay attention to recurring themes or feelings that arise, as they often point to areas that need attention or care.

3. Gratitude Practice: Reflecting on what you're grateful for nurtures the soul by shifting your focus from what's lacking to what's abundant in your life. Each day, take a moment to list three things you're grateful for. Over time, this practice fosters a mindset of appreciation and positivity.

"Gratitude turns what we have into enough." – Anonymous

By engaging in these practices, self-reflection becomes a daily ritual that nurtures your inner world and strengthens your connection with your authentic self. It creates space for growth, healing, and a deeper understanding of your purpose.

Spiritual Growth: Aligning with Your True Purpose

"The meaning of life is to find your gift. The purpose of life is to give it away." – Pablo Picasso

Spiritual growth is a journey of aligning your actions, values, and beliefs with your inner purpose. It's about seeking meaning beyond material achievements and connecting with something greater—whether that's your own potential, a higher power, or the beauty of the world around you. Cultivating spiritual growth brings a sense of fulfillment that resonates deeply within and provides clarity amid life's uncertainties.

One way to foster spiritual growth is by **engaging in meaningful rituals**. Rituals create space for intention and reflection, offering a moment to pause and connect with what matters most. These can be simple practices like praying, lighting a candle while meditating, taking a mindful walk in nature, or dedicating time each evening to reflect on the day's events. These rituals help ground you and remind you of your purpose.

Connection with nature is another powerful way to nurture the soul. Spending time outdoors—

whether walking through a park, hiking in the mountains, or simply sitting by a body of water—provides a sense of awe and perspective. Nature reminds us of life's rhythms and the interconnectedness of all things, fostering a sense of peace and wonder.

"Look deep into nature, and then you will understand everything better." – Albert Einstein

Spiritual growth also involves **cultivating compassion and service**. Acts of kindness, whether big or small, connect us with others and reinforce our shared humanity. Helping someone in need or contributing to a cause that aligns with your values not only enriches their life but also deepens your sense of purpose and connection.

By aligning your actions with your inner values and dedicating time to practices that promote reflection and connection, you can foster spiritual growth that nurtures both your soul and your sense of purpose.

Daily Practices to Nurture Your Soul: Creating a Routine of Inner Peace

"What lies behind us and what lies before us are tiny matters compared to what lies within us." – Ralph Waldo Emerson

Nurturing the soul doesn't require grand gestures or life-altering changes; it's about weaving small, intentional practices into your daily life that create moments of calm, connection, and self-awareness. By integrating these practices into your routine, you cultivate a sense of inner peace and alignment that sustains you through life's challenges.

1. Morning Mindfulness: Start your day with a few moments of mindfulness. Before reaching for your phone or diving into your to-do list, take five to ten minutes to sit quietly, center yourself by focusing on your breath, express gratitude for the present moment, and set a positive intention for the day. This practice creates a sense of grounding and helps you approach the day with clarity and purpose.

"Morning mindfulness brings clarity and peace to your day, helping you start with a calm mind and a centered heart." -Alex Salem

2. Creative Expression: Engage in activities that allow your soul to express itself freely, such as painting, writing, playing music, or crafting. Creative expression provides an outlet for emotions and thoughts that may be difficult to articulate otherwise, fostering a deeper connection with your inner self.

"Creativity thrives in connection, where ideas flow freely and inspiration is shared." -Alex Salem

3. Affirmations and Positive Self-Talk: Begin or end your day by repeating affirmations that resonate with your values and aspirations. Phrases like "I am enough," "I am capable of growth," or "I am at peace with myself" can help shift your mindset toward positivity and self-compassion.

"I embrace each day with positivity, knowing my creativity leads to amazing possibilities."-Alex Salem

4. Evening Reflection: Dedicate time at the end of the day to reflect on what went well, what you learned, and how you felt. This practice of gratitude and introspection helps you acknowledge progress and cultivate a sense of contentment.

By incorporating these practices into your daily life, you create a routine that nurtures your soul and reinforces your connection to your inner self. These small, consistent efforts have a profound impact on your spiritual well-being, helping you navigate life with balance and resilience.

"Your thoughts shape your reality. Choose them wisely." – Unknown

Integrating Spiritual Practices into a Modern Lifestyle

"In the midst of movement and chaos, keep stillness inside of you." – Deepak Chopra

In today's fast-paced, constantly connected world, it may seem challenging to prioritize spiritual well-being. However, spiritual growth doesn't require detaching from everyday responsibilities or spending hours in solitude. Instead, it's about creating meaningful moments of connection, reflection, and purpose within the flow of your modern life. By weaving spiritual practices into your routine, you can nurture your soul while balancing personal and professional demands.

One of the most effective ways to integrate spiritual practices is through **mindful multitasking**. This doesn't mean doing more things at once but rather bringing presence and intention to activities you're already doing. For example:

- **Mindful Commuting:** Use your commute as a time for reflection or gratitude rather than filling it with distractions. Consider what you're looking forward to in the day ahead, or reflect on what you're thankful for in your life.
- **Intentional Breaks:** During the workday, take a few moments to pause and breathe deeply. Even a short mental reset can help you reconnect with your inner self and refocus your energy.
- **Household Chores with Purpose:** While cooking or cleaning, focus on the sensations and movements involved, using the activity as a form of meditation. This practice transforms mundane tasks into moments of presence.

Another way to foster spiritual well-being is to **build community**. Spiritual growth often flourishes in connection with others, whether through shared values, supportive relationships, or acts of kindness. Joining a group that aligns with your beliefs—whether it's a meditation circle, a volunteer organization, or a creative workshop—provides opportunities for growth, inspiration, and belonging. Community nurtures the soul by reinforcing the idea that we are all part of something greater than ourselves.

"We rise by lifting others." – Robert Ingersoll

Additionally, **regularly disconnecting from technology** is essential for maintaining spiritual balance. Spending too much time in the digital world can leave us feeling disconnected from ourselves. Set boundaries for screen time, and dedicate at least part of each day to unplugging—whether it's during meals, an evening walk, or a reflective morning routine. These moments of disconnection create space for self-reflection and mindfulness.

Here are some practical steps to help you disconnect from technology and create a healthier balance in your life:

Set Specific Time Limits: Allocate specific time slots for technology use (e.g., social media, emails, streaming). Stick to these limits and avoid mindless scrolling.

Turn Off Notifications: Disable non-essential notifications on your devices. This reduces distractions and helps you stay focused.

Designate Tech-Free Zones: Create areas in your home where technology is not allowed (e.g., the bedroom during sleeping hours).

Unplug During Meals: Avoid using devices while eating. Instead, savor your food and engage in conversation.

Go for a Walk without your phone: Step outside and enjoy nature. Walking without distractions allows you to clear your mind.

Read a Physical Book: Put away digital devices and read a physical book. It's a great way to disconnect and engage your mind.

Schedule Screen-Free Time: Set aside specific hours each day for screen-free activities (e.g., hobbies, exercise, meditation).

Use Grayscale Mode: Temporarily switch your phone to grayscale mode. This makes apps less appealing and reduces screen time.

Practice Digital Detox Days: Dedicate entire days to disconnecting from technology. Use this time for self-reflection, creativity, or outdoor activities.

Keep in mind that even small changes can help you build a healthier relationship with technology!

Finally, consider creating a **sacred space** in your home where you can retreat for quiet reflection or meditation. This space doesn't have to be elaborate; it can be as simple as a corner with a comfortable chair, a candle, and a few meaningful objects. This dedicated space serves as a reminder to prioritize your spiritual well-being, offering a place to recharge and realign.

"Your sacred space is where you can find yourself again and again." – Joseph Campbell

By integrating these practices into your life, you create a sustainable approach to spiritual growth that complements your lifestyle. Nurturing the soul is not about perfection but about progress—a continuous journey of connection, reflection, and purpose. Over time, these small, intentional acts build a foundation for a life that feels balanced, fulfilling, and deeply aligned with your inner self.

Chapter 17: Financial Strategies

The Foundations of Financial Security: A Path to Stability and Growth

"In business, I've discovered that my purpose is to do my best to my utmost ability every day. That's my standard." – President Donald Trump

In an era of rapid economic changes, marked by AI-driven workforce disruptions, rising debts, and inflationary pressures, financial security has become a cornerstone of personal well-being. Financial strategies are no longer optional—they are essential tools for navigating an uncertain economy, building stability, and securing a prosperous future. This chapter aims to equip readers with practical strategies to improve their financial resilience, minimize losses, and prepare for economic uncertainties.

Financial security begins with understanding **your current financial situation**. Before creating strategies for improvement, take the time to evaluate your income, expenses, debts, and savings. This assessment provides a clear picture of where you stand financially and identifies areas that require attention. Tools like budgeting apps, financial trackers, or even simple spreadsheets can help you gain insight into your financial habits, revealing patterns that either support or undermine your goals.

A key principle of financial security is **establishing an emergency fund**. An emergency fund is a savings buffer designed to cover unexpected expenses, such as medical bills, car repairs, or temporary job loss. Financial experts recommend setting aside at least three to six months' worth of living expenses. This fund acts as a safety net, preventing reliance on credit cards or loans during difficult times and reducing stress in moments of crisis.

"A big part of financial freedom is having your heart and mind free from worry about the what-ifs of life." – Suze Orman

This chapter will also explore strategies for **budgeting, investing, and diversifying income streams**, helping readers create a financial plan that is both robust and adaptable. By mastering these principles, you can gain greater control over your finances, protect yourself from economic instability, and pave the way for long-term growth.

Budgeting: The Cornerstone of Financial Management

"A budget is telling your money where to go instead of wondering where it went." – Dave Ramsey

Budgeting is one of the most powerful tools for achieving financial security. It provides a clear framework for managing your income, controlling expenses, and saving for future goals. Far from being restrictive, a well-designed budget offers freedom. It allows you to make informed decisions about your finances and ensures that your money is working toward your priorities.

"Financial freedom is available to those who learn about it and work for it." – Robert Kiyosaki

Vital Steps for Mastering Finance and Budgeting:

Achieving long-term success and financial stability requires a solid understanding of money and budgeting. You can take charge of your financial destiny and make wise decisions that support your objectives if you know where your money is coming from and going. In addition to helping you pay for everyday costs, a well-structured budget also enables you to plan for the future, save for emergencies, and properly enjoy life's blessings. The stages listed below offer a clear road map to assist you in managing your money, staying on course, and laying the groundwork for financial independence. But first, let's explore what you need to know before creating a budget for yourself.

Income: List all sources of income (e.g., salary, freelance work, investments) and include the amount you receive from each source.

Expenses: Categorize your expenses (e.g., housing, utilities, groceries, transportation). Estimate the monthly cost for each category.

Fixed Expenses: These are regular, predictable expenses (e.g., rent/mortgage, loan payments). Allocate funds for fixed expenses first.

Variable Expenses: These fluctuate month-to-month (e.g., dining out, entertainment). Set limits for variable expenses.

Emergency Fund: Allocate a portion of your income to an emergency fund. Aim for 3-6 months' worth of living expenses.

Savings Goals: Identify specific financial goals (e.g., vacation, retirement). Allocate funds toward these goals.

Review and Adjust: Regularly review your budget. Adjust as needed based on changes in income or expenses.

Remember, the key to a successful budget is consistency and tracking your spending!

"Embrace the Power of Budgeting: You're Crafting Your Future Canvas." – Alex Salem

Now, let's learn how to create and plan an effective budget for yourself.

"When you set a budget, you are taking control of your future."

1. **Start with a Simple Budget Plan:** A basic budgeting method, such as the **50/20/10/10/10 rule**, is an excellent starting point. Allocate:

- **Necessary Expenses (50%):** Allocate 50% of your income for essential needs. Includes housing (rent or mortgage), utilities, groceries, transportation, and insurance.
- **Financial Goals (20%):** Set aside 20% for savings, investments, and debt payments. Prioritize building an emergency fund, contributing to retirement accounts, and paying off debt.
- **Discretionary Spending (10%):** Use 10% for non-essential expenses. Includes entertainment, dining out, hobbies, and personal indulgences.
- **Giving or Charity (10%):** Dedicate 10% to charitable donations or supporting causes you care about.
- **Education or Personal Growth (10%):** Allocate 10% for personal growth: Invest in learning, courses, and books. Prioritize yourself by paying yourself first!

This structure creates balance, ensuring that your needs are met while leaving room for enjoyment and future planning.

2. Track Your Spending: Monitoring your expenses is critical for maintaining control over your budget. Use financial apps or spreadsheets to categorize and track where your money goes. Identifying patterns helps you spot areas where you may be overspending and allows you to redirect those funds toward savings or investments.

3. Prioritize Savings: Automating your savings ensures that they are not left to chance. Set up automatic transfers to your savings account as soon as you receive your income. Treating savings like a fixed expense creates consistency and builds financial security over time.

"Flexibility is the key to adapting to shifts without losing sight of our goals. Evolve, not dissolve" - Alex Salem

4. Adjust Your Budget Regularly: Economic circumstances, income changes, and evolving goals may require adjustments to your budget. Review it monthly to ensure it aligns with your current needs and aspirations. Flexibility is key to adapting to financial shifts without losing sight of your objectives.

Budgeting is not about deprivation. It's about intention. By creating a plan for your money, you gain clarity, control, and confidence, enabling you to meet immediate needs while building a secure financial future.

Investing: Building Wealth and Securing the Future

"Do something today that your future self will thank you for." – Sean Patrick Flanery

Investing is a critical component of financial strategies, transforming saved income into wealth that grows over time. With inflation constantly eroding the value of cash, investing allows individuals to outpace inflation, build wealth, and secure long-term financial goals. However, successful investing requires knowledge, planning, and the ability to manage risk.

"An investment in knowledge pays the best interest." - Benjamin Franklin

1. Understanding Investment Options: Before diving into investing, it's essential to understand the various avenues available:

- **Stocks:** Represent shares in a company, offering potentially high returns but with increased volatility.
- **Bonds:** Debt securities issued by governments or corporations, providing more stable but lower returns.
- **Mutual Funds and ETFs:** These are pooled investment options that spread risk by investing in a diversified portfolio of stocks, bonds, or other securities.

- **Real Estate:** Physical property investments that offer long-term appreciation and rental income.
- **Cryptocurrencies and Emerging Markets:** Riskier, speculative investments that can yield high rewards but require caution.

Each option comes with its own risk-reward profile, and the best choice depends on your financial goals, risk tolerance, and investment timeline.

2. Start Small but Start Early: The earlier you begin investing, the more time your money has to benefit from **compound interest**, the process where your earnings generate more earnings over time. Even small, consistent investments can grow substantially when compounded over decades. For example, investing $100 monthly in a diversified index fund with an average annual return of 7% can grow to over $120,000 in 30 years.

"Big things have small beginnings." – Prometheus

3. Diversify to Manage Risk: One of the golden rules of investing is diversification—spreading investments across different asset classes and industries to minimize the impact of any single loss. For instance, if one sector underperforms, gains in another can balance your overall portfolio.

"The key to risk management is never putting yourself in a position where you cannot live to fight another day." - Richard S. Fuld, Jr.

4. Educate Yourself Continuously: Understanding market trends, financial products, and investment strategies is crucial for making informed decisions. Free resources, such as financial podcasts, books, and webinars, provide valuable insights. For more personalized advice, consider consulting a financial advisor who can tailor an investment plan to your unique needs and goals.

"The stronger your mindset is, the greater your skill set will be. The mind has no limitations—nourish and challenge it daily. Learn, teach, and educate!" -Alex Salem

5. Stay Patient and Disciplined: The stock market and other investments often experience short-term fluctuations. A long-term perspective helps investors avoid emotional decisions during market downturns. Regularly review your investment portfolio and rebalance it if necessary, but avoid impulsive actions that may harm your financial growth.

Investing is not a get-rich-quick strategy—it's a disciplined, long-term approach to building wealth. By starting early, staying informed, and managing risk effectively, you can create a portfolio that supports financial security and independence.

"Patience is choosing between what you want now and what you want most." – Abraham Lincoln

Diversifying Income Streams: Adapting to an Evolving Economy

"Income diversification is the key to financial resilience." – Suze Orman

In a world where technological advancements, particularly **AI**, are reshaping industries and job markets, relying solely on one source of income is increasingly risky. Diversifying your income streams not only ensures financial stability but also creates opportunities for growth and security in uncertain times. This approach allows you to spread financial risk across multiple avenues, reducing the impact of any one source drying up.

"Great things come from hard work and perseverance. No excuses." – Kobe Bryant

1. Side Hustles and Freelancing: Unlocking Flexibility

The rise of the gig economy has opened countless opportunities to earn additional income. Side hustles such as freelancing, consulting, or selling products online allow individuals to monetize their skills or passions. For example:

- A graphic designer might offer freelance services through platforms like Fiverr or Upwork.
- A teacher could tutor online or create educational courses.
- A writer might monetize a blog or publish e-books.

These pursuits offer flexibility, allowing you to manage your primary job while earning supplemental income. Additionally, side hustles often develop into full-fledged businesses over time, opening doors to greater financial independence.

2. Passive Income: Earning While You Sleep

Creating streams of passive income is a powerful way to build wealth without continuous active involvement. Passive income can come from:

- **Investments:** Stocks, bonds, or real estate generate dividends, interest, or rental income.
- **Royalties:** Selling intellectual property, such as books, music, or software, provides ongoing revenue.
- **Digital Products:** Courses, templates, or apps can generate income long after they're created.

While these streams may require upfront effort or capital, they offer long-term rewards, freeing you from relying solely on active income.

"If you don't find a way to make money while you sleep, you will work until you die." – Warren Buffett

3. Upskilling for Economic Adaptability

As **AI** and automation disrupt traditional jobs, staying competitive requires **continuous learning and skill development**. Identify skills in demand within your industry or adjacent sectors and invest in acquiring them. For instance:

- A marketer might learn **AI-driven** analytics tools to stay relevant in a data-centric job market.
- A software developer could expand into **AI** and machine learning programming.
- Professionals in creative fields might adopt **AI-powered** design or writing tools to enhance their work.

Upskilling not only safeguards your primary income but also opens new streams of revenue by allowing you to branch out into related roles or freelance opportunities.

"Intelligence is the ability to adapt to change." - Stephen Hawking

4. Small Business Ventures: Turning Ideas into Income

"Entrepreneurship is the art of transforming your passion into a thriving business." – Alex Salem

Entrepreneurship offers the chance to diversify income while pursuing something personally meaningful. Whether it's starting an online store, providing a niche service, or launching a passion project, small business ventures can grow into substantial income sources over time. Platforms like Etsy, Shopify, and Amazon simplify the process of reaching customers worldwide.

"Opportunities don't happen. You create them." – Chris Grosser

5. Adapting to Economic Shifts

Diversifying income also means staying alert to economic trends and adjusting accordingly. For example, during inflationary periods, investing in assets like real estate or gold can protect wealth. Similarly, adapting to technological trends—like creating content for emerging platforms or learning about blockchain—ensures you remain ahead of the curve.

By cultivating multiple income streams, you safeguard your financial future and reduce reliance on a single source. This strategy not only creates resilience but also empowers you to seize new opportunities, achieve financial freedom, and thrive in an ever-changing economy.

Planning for Unforeseen Circumstances: Building Resilience in Financial Uncertainty

"By failing to prepare, you are preparing to fail." – Benjamin Franklin

One of the most crucial aspects of financial security is preparing for the unexpected. Economic instability, medical emergencies, or unforeseen personal challenges can disrupt even the most carefully constructed financial plans. By proactively addressing these uncertainties, you can minimize their impact and maintain financial stability in the face of adversity.

1. Building and Maintaining an Emergency Fund

An emergency fund acts as your first line of defense against financial disruptions. As noted earlier, this fund should ideally cover three to six months' worth of essential living expenses. Here's how to build and maintain it:

- Start small: Aim for a target amount (e.g., one month's expenses) and grow it over time.
- Automate savings: Set up automatic transfers to your emergency fund with each paycheck.

- Keep it accessible but separate: Use a high-yield savings account to earn interest while ensuring the funds are readily available.

This fund ensures you can handle unexpected events, like job loss, car repairs, or divorce, without relying on credit cards or loans.

2. Insurance: Protecting Against Risk

Insurance is another vital tool for managing unforeseen circumstances. Key types of insurance to consider include:

- **Health Insurance:** Covers medical expenses that could otherwise lead to significant debt.
- **Life Insurance:** Provides financial security for your family in the event of your passing.
- **Disability Insurance:** Replaces income if you're unable to work due to injury or illness.
- **Property Insurance:** Protects assets like your home or car.

Reviewing your insurance coverage regularly ensures it remains adequate for your evolving needs.

3. Creating a Financial Plan for Emergencies

Developing a contingency plan helps you act decisively during a crisis. This plan might include:

- A prioritized list of expenses to cut if necessary.
- A strategy for accessing funds (e.g., emergency savings, borrowing options).
- A checklist of key financial documents, such as insurance policies, bank account information, and emergency contacts.

Having a plan reduces stress and ensures that you're prepared to handle challenges with confidence.

4. Preparing for Economic Instability

Economic instability, such as recessions or inflation, can affect everyone. To safeguard your finances:

- **Diversify investments:** Spreading investments across asset classes (stocks, bonds, real estate) reduces exposure to market volatility.
- **Maintain liquidity:** Ensure part of your portfolio remains liquid to address unexpected needs.
- **Stay informed:** Monitor economic trends and adjust your financial strategies accordingly, such as reallocating assets during inflationary periods or seeking lower-risk investments during recessions.

"Every decision you make today shapes your tomorrow. Prepare with purpose, not just repair with regret." - Alex Salem

5. Long-Term Financial Planning

Preparing for the future also involves creating a robust retirement plan. Contributing consistently to retirement accounts (e.g., 401(k), IRA) ensures you're building a safety net for later years. Take advantage of employer matching programs and tax-advantaged accounts to maximize savings.

Integrating Financial Security into Daily Life

Financial resilience isn't about living in fear of the future but about empowering yourself with preparation and adaptability. By incorporating financial planning into your routine—whether through budgeting check-ins, regular investment reviews, or consistent contributions to savings—you create a proactive approach to long-term stability.

"The question isn't at what age I want to retire; it's at what income." – George Foreman

In conclusion, planning for unforeseen circumstances requires foresight, discipline, and flexibility. By building an emergency fund, securing insurance, and preparing for economic instability, you can create a financial strategy that protects you and your loved ones in the face of life's uncertainties.

An important point to remember: The journey to success isn't just about accumulating wealth; it's about becoming the best version of yourself. It's about developing the qualities and mindset required to achieve your goals. In essence, becoming a "self-made millionaire" isn't merely about amassing wealth; it's about transforming into a different person — one who embodies discipline, determination, resilience, and a relentless pursuit of excellence on the path to success.

Chapter 18: Navigating Work

The Changing Landscape of Work: How AI is Reshaping the Workplace

"The future of work isn't about man versus machine. It's about man with machine." – Satya Nadella

Artificial intelligence is fundamentally transforming the nature of work, introducing both exciting opportunities and significant challenges. As **AI** continues to evolve, it reshapes industries, redefines job roles, and creates a new paradigm of employment dynamics. While AI has the potential to boost efficiency and innovation, it also raises concerns about job displacement and the skills gap, leaving many workers wondering how to navigate this rapidly changing landscape.

AI's impact is especially visible in industries that rely on repetitive tasks and data processing. For example:

- In **manufacturing**, AI-driven robotics automate production lines, reducing costs and increasing precision. However, this efficiency often comes at the expense of human roles traditionally tied to manual labor.

- In **finance**, algorithms perform complex calculations and analyses in seconds, replacing the need for large teams of analysts for routine tasks like fraud detection or investment predictions.
- In **healthcare**, AI enhances diagnostic accuracy and streamlines administrative tasks, allowing medical professionals to focus more on patient care.

Despite these advances, concerns about job security loom large. A report by the World Economic Forum estimated that AI and automation could displace **85 million jobs worldwide by 2025** while simultaneously creating **97 million new roles** requiring advanced skills. This duality highlights the need for adaptability and continuous learning in today's workforce.

"AI-driven automation will continue to eliminate traditional jobs, but also create new opportunities and industries we can't even imagine yet." - Jensen Huang

AI is not just eliminating jobs—it's creating new opportunities. Roles like **AI specialists, data scientists, machine learning engineers, and human-AI interaction designers** have emerged, requiring a blend of technical expertise and creative problem-solving. The rise of AI also emphasizes the need for uniquely human skills, such as emotional intelligence, critical thinking, and leadership.

"Artificial intelligence doesn't eliminate jobs, ignoring it does. Embrace AI to innovate, evolve, and create new growth opportunities."- Alex Salem

This section explores the evolving nature of work, highlighting how AI is both a challenge and an ally. By understanding its impact, we can better prepare for a future where collaboration between humans and machines defines workplace success.

Education and Professions: Preparing for AI-Driven Transformations

"The future of AI is bright. By working together, humans and AI can create a world of unimaginable possibilities." - Jean-Claude Tremblay

As **AI** reshapes industries, it also demands significant changes in education systems and professional training. Traditional learning models, which often emphasize rote memorization and standardized testing, are increasingly inadequate for preparing students and workers to succeed in an AI-driven world. The rapid pace of technological advancements requires an education system that emphasizes adaptability, creativity, and lifelong learning.

1. The Role of Education in the AI Era

Education systems must pivot from knowledge transfer to **skill development**. AI's rise underscores the importance of teaching foundational skills like coding, data analysis, and digital literacy. However, equally critical are **soft skills** such as emotional intelligence, teamwork, and communication, which remain uniquely human strengths.

AI-powered tools are also transforming how students learn. Adaptive learning platforms, for example, use AI to personalize educational content based on a student's strengths and weaknesses. This allows for tailored instruction that maximizes individual potential. Yet, these tools are only effective when combined with mentorship and guidance from educators who foster creativity and critical thinking.

2. Impacts on Specific Professions

AI affects every profession differently, creating both disruption and opportunities:

- **Healthcare:** AI streamlines administrative tasks, assists in diagnostics, and predicts patient outcomes. This enables healthcare professionals to spend more time with patients, but it also necessitates training to work alongside AI systems.
- **Legal Services:** AI tools expedite legal research, contract review, and case predictions, reducing the workload for paralegals and junior attorneys. Lawyers now need to focus more on strategy and client relations.

- **Creative Industries:** AI tools like generative design and content creation software enhance creativity but challenge traditional notions of originality. Professionals in these fields must learn to leverage AI as a collaborator rather than a competitor.
- **Logistics and Transportation:** Autonomous vehicles and AI-driven route optimization disrupt traditional roles, requiring workers to upskill or shift to supervisory and maintenance positions.

"In the Era of AI, success depends not just on what you know but on how you think and adapt. Accept the future with an open mind and a courageous heart." - Alex Salem

3. Lifelong Learning and Reskilling

The era of AI demands a shift toward **lifelong learning**. Workers must regularly update their skills to stay relevant in dynamic industries. Online platforms like Coursera, Udemy, and LinkedIn Learning offer accessible courses in AI-related skills, making it easier for individuals to reskill and adapt to industry demands.

"Learn continually—there's always one more thing to learn." – Steve Jobs

AI is not just altering how we work but how we prepare for work. Education systems and professionals must embrace these changes to remain agile and competitive in an ever-evolving economy.

Industry-Specific Impacts: Opportunities and Challenges in an AI-Driven Economy

"How you perceive transformation determines your path: see it as an obstacle, and you'll struggle; see it as an opportunity, and you'll thrive."- Alex Salem

Artificial intelligence is transforming industries across the board, creating a ripple effect of opportunities and challenges. By understanding these industry-specific impacts, workers and employers alike can anticipate changes, adapt to new roles, and capitalize on emerging possibilities.

1. Manufacturing: Automation at Scale

AI and robotics have revolutionized manufacturing, bringing unparalleled efficiency and precision. Factories now use AI-driven robots to assemble products, monitor quality, and optimize supply chains. While these advancements reduce costs and boost productivity, they also displace roles traditionally held by human workers, such as assembly line operators.

- **Challenges:** Workers without technical skills risk losing jobs to automation. Companies face pressure to reskill employees and manage the human impact of technological shifts.

- **Opportunities:** New roles in programming, maintaining, and supervising automated systems are emerging. Workers trained in mechatronics, industrial AI, or robotics have a competitive edge in this evolving field.

"AI-powered robots will become a common sight in our daily lives, performing tasks that humans once thought impossible for machines." - Hiroaki Kitano

2. Healthcare: Enhanced Diagnostics and Patient Care

AI in healthcare offers life-saving innovations, from predictive diagnostics to personalized medicine. For instance, AI algorithms can analyze medical imaging to detect diseases like cancer with higher accuracy than traditional methods. It also supports wearable devices that monitor patient health in real-time.

- **Challenges:** The adoption of AI raises ethical concerns about data privacy and decision-making. Professionals must adapt to working alongside AI without losing the human touch critical to patient care.
- **Opportunities:** AI creates demand for healthcare IT specialists, data analysts, and professionals skilled in integrating AI into medical workflows.

"AI has allowed me, as a physician, to be 100% present for my patients." - Michelle Thompson

3. Finance: Transforming Analytics and Customer Service

The financial sector has embraced AI for fraud detection, risk management, and customer service. Chatbots and robo-advisors have streamlined client interactions, while machine learning models analyze massive datasets to predict market trends and manage investments.

- **Challenges:** AI's efficiency in areas like data analysis reduces demand for entry-level roles, such as bank tellers or financial assistants.
- **Opportunities:** Roles in financial data science, algorithmic trading, and AI ethics oversight are becoming increasingly valuable.

"AI in customer service can provide personalized experiences at scale, something that was previously impossible." - Marc Benioff

4. Creative Industries: Augmentation, Not Replacement

AI tools like GPT-4, GPT-4o, DALL·E, and Adobe's AI-driven software are reshaping creative professions. Artists, writers, and designers can now use AI to generate ideas, refine concepts, and automate repetitive tasks.

- **Challenges:** Concerns about originality and the devaluation of human creativity persist. Professionals may feel pressure to prove their relevance in an AI-augmented landscape.
- **Opportunities:** Creatives who learn to integrate AI into their workflow can amplify their productivity and reach. Roles in AI-assisted design and multimedia storytelling are expanding.

"AI is about augmenting human intelligence, not replacing it." - Ginni Rometty

5. Logistics and Transportation: Automation in Motion

AI's role in logistics includes optimizing delivery routes, predicting maintenance needs, and enabling autonomous vehicles. Companies like **Amazon** and **Tesla** are at the forefront of deploying AI to streamline supply chains and transportation networks.

- **Challenges:** Drivers and warehouse workers face job displacement as automation takes over routine tasks.
- **Opportunities:** New roles in AI-driven logistics planning, maintenance of autonomous systems, and data analysis

emerge as critical components of the industry.

"How to harness adaptability to transform chaos from an obstacle into an opportunity." – Alex Salem

AI's impact varies across industries, but the common thread is adaptability. Those who stay informed, upskill and embrace AI as a partner rather than a threat will thrive in the evolving job market.

"AI and automation are the engines of the next industrial revolution, transforming industries and creating new possibilities." - Elon Musk

Adapting to New Job Opportunities: Thriving in the AI Era

"It is not the strongest of the species that survives, nor the most intelligent, but the one most responsive to change." – Charles Darwin

As **AI** continues to disrupt industries, adaptation becomes the cornerstone of professional success. The emergence of AI technologies is creating new job opportunities in areas that didn't exist a decade ago while also demanding that workers develop skills to remain competitive. To navigate this landscape, individuals must embrace a proactive approach, focusing on reskilling, networking, and adopting a growth mindset.

1. Reskilling and Upskilling for AI-Driven Roles

Reskilling involves acquiring entirely new skills for a different role, while upskilling focuses on enhancing your existing skill set to remain competitive in your field. Both are crucial for thriving in the AI-driven economy. Here's how to approach them:

- **Identify Trends:** Research industry shifts to determine which skills are becoming essential. For example, knowledge of machine learning and programming languages like Python or cloud computing is increasingly in demand across sectors.

- **Leverage Online Learning Platforms:** Platforms like Coursera, Udemy, and edX offer affordable courses in AI-related fields, such as data analysis, digital marketing, and cybersecurity.

- **Seek Employer-Supported Training:** Many organizations are now offering training programs to help employees transition into new roles. Taking advantage of these resources demonstrates initiative and positions you for growth.

"We cannot solve our problems with the same thinking we used when we created them." – Albert Einstein

2. Embracing Transferable Skills

While technical expertise is valuable, transferable skills—those applicable across various roles and industries—are equally critical. Examples include:

- **Problem-Solving:** Analyzing complex issues and finding innovative solutions is a skill AI cannot fully replicate.
- **Adaptability:** Remaining flexible in the face of change ensures you can transition seamlessly between roles.
- **Emotional Intelligence:** Understanding and managing emotions, both your own and others, is increasingly valued as AI handles routine tasks.

These skills provide stability in an evolving workplace, ensuring you remain relevant regardless of specific technological changes.

3. Networking in the AI-Driven World

Building a strong professional network is vital for staying informed and discovering new opportunities. Networking allows you to:

- **Learn About Trends:** Engaging with professionals in your industry helps you stay updated on technological shifts and emerging roles.
- **Access Hidden Opportunities:** Many roles are filled with recommendations before they're publicly advertised.
- **Collaborate on Projects:** Networking fosters partnerships that may lead to innovative projects, expanding your expertise and portfolio.

Use platforms like LinkedIn to connect with experts in AI, attend webinars or conferences, and participate in community events related to your field.

4. Embracing a Growth Mindset

A growth mindset—believing that abilities can be developed through dedication, transformation, and determination is essential for navigating AI-driven changes. This mindset enables individuals to:

- View challenges as opportunities for growth.
- Persist through setbacks and continue learning.
- Stay open to new technologies and methodologies.

By approaching the AI-driven workplace with curiosity, adaptability, and a commitment to learning, workers can not only survive but thrive.

The key lies in viewing AI as a supportive *tool* for enhancement rather than competition and continuously evolving alongside the technologies shaping the modern job market.

"Commitment is what transforms a promise into reality." – Abraham Lincoln

Strategies for Managing Challenges and Seizing Opportunities

"Every success story is a tale of constant adaptation, revision, and change." – Richard Branson

While **artificial intelligence** has brought remarkable opportunities, it also presents unique challenges that require strategic management. To navigate the evolving workplace landscape effectively, individuals must adopt a proactive approach, balancing the need for preparation with the ability to embrace the opportunities that AI creates. This section explores actionable strategies to manage these challenges while capitalizing on AI's potential.

1. Proactively Addressing Job Displacement

One of the most significant concerns about AI is the potential for job displacement. To stay ahead:

- **Stay Informed:** Follow industry news and trends to anticipate how AI might affect your field. Knowledge is your first line of defense.

- **Adapt Early:** If you notice automation taking over specific roles, start reskilling immediately. Workers who adapt early often transition more smoothly into new positions.
- **Engage in Cross-Training:** Develop skills outside your primary role. For example, a marketing professional might train in data analytics or user experience design, enhancing versatility.

"Success is not final, failure is not fatal: It is the courage to continue that count." – Winston Churchill

2. Focusing on Human-Centric Roles

AI excels at routine and data-intensive tasks, but it cannot replicate human creativity, empathy, or critical thinking. Focus on roles that require:

- **Emotional Intelligence:** Fields like counseling, human resources, and leadership demand interpersonal skills that AI cannot match.
- **Creativity and Innovation:** Content creation, product design, and storytelling remain deeply human endeavors.
- **Strategic Thinking:** Jobs requiring nuanced decision-making and long-term

planning will continue to rely on human expertise.

By prioritizing roles where humans outperform machines, your future-proof your career while embracing areas where AI complements rather than replaces.

"The greatest partnership is between humans and machines—united, we unlock limitless possibilities." – Alex Salem

3. Learning to Work Alongside AI

Collaboration with AI is quickly becoming a critical workplace skill. To harness its potential:

- **Understand AI Tools:** Familiarize yourself with AI applications relevant to your field. For instance, a graphic designer might learn AI-powered design software, while a teacher could explore adaptive learning platforms.

- **Leverage AI for Efficiency:** Use AI to automate repetitive tasks, freeing time for strategic and creative work.

- **Develop a Partnership Mindset:** Treat AI as a collaborator. For example, in journalism, AI can analyze data or generate drafts, but human writers add depth and context.

4. Advocating for Ethical AI Adoption

AI's implementation raises ethical concerns, including bias, data privacy, and accountability. Workers and leaders alike must advocate for responsible AI usage by:

- **Demanding Transparency:** Ensure AI systems used in your workplace operate fairly and explain their decision-making processes.
- **Participating in Policy Development:** If possible, contribute to discussions around how AI is adopted in your organization, emphasizing fairness and inclusiveness.
- **Educating Yourself on AI Ethics:** Understanding the ethical dimensions of AI empowers you to advocate for practices that benefit both organizations and workers.

5. Embracing the Opportunities of a Reshaped Economy

AI's transformative power creates a wealth of new opportunities:

- **Entrepreneurship:** Many startups are leveraging AI to innovate in healthcare, education, and logistics. Explore how you can harness AI to address unmet needs.

- **Remote Work Expansion:** AI tools make remote work more effective, opening doors to global job opportunities.
- **Customized Career Paths:** With AI automating routine tasks, professionals have more time to pursue specialized interests and build careers aligned with their passions.

"Opportunities are usually disguised as hard work, so most people don't recognize them." – Ann Landers

Conclusion

Navigating the AI-driven workplace requires a mix of adaptability, resilience, and strategic thinking. By proactively addressing challenges, focusing on unique human skills, and embracing collaboration with AI, you position yourself for success in a rapidly changing world. The key is to view **AI** not as a threat but as a catalyst for innovation and growth, unlocking new possibilities for personal and professional achievement.

Chapter 19: Building Meaningful Relationships

The Value of Meaningful Relationships: Human Connection in a Changing World

"Connection is why we're here; it is what gives purpose and meaning to our lives." – Brené Brown

The rise of artificial intelligence, the pervasive influence of social media, and the societal shifts following the COVID-19 pandemic have profoundly impacted how we connect with one another. While technology has made communication more convenient, it has also led to an increase in isolation and superficial interactions. The illusion of connection—whether through AI-driven chatbots or curated social media feeds—often leaves people feeling more disconnected than ever before.

"The only way to have a friend is to be one." - Ralph Waldo Emerson

Meaningful relationships are vital for emotional well-being and personal growth. They provide a sense of belonging, reduce stress, and foster resilience during life's challenges. Yet, the comfort of isolation, fueled by digital interactions, has made building and maintaining genuine connections increasingly difficult. Loneliness is now considered a global epidemic, with research showing its impact on health is comparable to smoking or obesity.

The Illusion of AI Companionship AI technologies, like conversational bots or virtual assistants, are designed to mimic human interaction, offering users the feeling of companionship. While these tools may provide temporary relief from loneliness, they lack the depth and empathy required for genuine connection. Over-reliance on AI for emotional support can deepen isolation, creating a barrier to forming real relationships.

"AI companionship sparkles, but it lacks the soul of true human connection." - Alex Salem

Social Media and Superficial Interaction Similarly, social media platforms often replace meaningful interactions with likes, comments, and fleeting messages. While these engagements create the illusion of connection, they rarely provide the emotional depth necessary for fulfilling relationships. Studies reveal that heavy social media usage is associated with higher rates of loneliness and depression, underscoring the need for intentional, face-to-face connections.

"Technology can create the illusion of companionship without the demands of friendship." – Sherry Turkle

This chapter explores the importance of genuine relationships and offers strategies to overcome the challenges of isolation in a digital world. By prioritizing meaningful connections, we can nurture our emotional well-being and build a network of support that enriches our lives.

Overcoming Isolation: Reconnecting in a Post-Pandemic World

"We are all so much together, but we are all dying of loneliness." – Albert Schweitzer

The COVID-19 pandemic disrupted traditional social norms, forcing people into isolation and redefining how we interact. Even as the world transitions back to normalcy, many individuals have struggled to regain the confidence and habits required to maintain personal relationships. The lingering effects of social distancing and a greater reliance on technology have made it harder for some to initiate and nurture meaningful connections.

1. Rebuilding Social Confidence

After prolonged periods of isolation, social interaction can feel daunting. Many people report heightened social anxiety or a sense of awkwardness in face-to-face encounters. Overcoming this requires gradual exposure and practice:

- **Start Small:** Begin with low-pressure social activities, such as meeting one trusted friend for coffee or attending a casual gathering.
- **Acknowledge Vulnerability:** It's normal to feel nervous when re-entering social settings. Share your feelings openly with friends or family, as vulnerability fosters connection.
- **Set Realistic Expectations:** Building or rekindling relationships takes time. Focus on enjoying the process rather than rushing to achieve a deep connection.

2. Addressing Emotional Isolation

Even when surrounded by people, emotional isolation—feeling disconnected or misunderstood—can persist. Combat this by fostering intentional communication:

- **Ask Open-Ended Questions:** Show genuine interest in others by asking questions like, "What's been on your mind lately?" or "What's something you're excited about?"
- **Practice Active Listening:** Be fully present in conversations, avoiding distractions like phones. Reflect back on what the other person shares to ensure they feel heard and valued.
- **Share Authentically:** Deep relationships require reciprocity. Be open about your

thoughts, experiences, and challenges to encourage mutual trust.

"Loneliness and the feeling of being unwanted is the most terrible poverty." – Mother Teresa

3. Transitioning from Virtual to In-Person Connections

While virtual communication was a lifeline during the pandemic, face-to-face interaction remains essential for building deeper bonds. Physical presence strengthens emotional connections, allowing us to pick up on nonverbal cues like body language and tone of voice. Transitioning to in-person connections can include:

- Hosting small gatherings at home.
- Joining local clubs or groups that align with your interests.
- Volunteering for community initiatives fosters connection while contributing to a shared cause.

By gradually overcoming isolation and reintroducing ourselves to meaningful social interactions, we rebuild the confidence and emotional resilience needed for lasting relationships. These steps lay the foundation for a more connected, fulfilling life in the post-pandemic world.

Cultivating Genuine Connections: Depth Over Quantity

"It is not the number of friends you have, but the depth of the friendships you possess." – Unknown

In a world where social media often emphasizes the number of followers, friends, or connections, the importance of deep, genuine relationships is often overlooked. Meaningful relationships prioritize quality over quantity, offering emotional support, trust, and mutual understanding that superficial interactions cannot provide.

---

1. Building Emotional Intimacy

Emotional intimacy is the cornerstone of genuine connections. It involves the ability to share your thoughts, feelings, and vulnerabilities without fear of judgment. To cultivate emotional intimacy:

- **Practice Empathy:** Make a conscious effort to understand others' emotions and perspectives. Reflect on how their experiences shape their worldview.
- **Express Gratitude:** Regularly acknowledge the value someone brings to your life. A simple "Thank you for being there for me" strengthens bonds.
- **Be Vulnerable:** Share your struggles and fears openly. Vulnerability fosters trust and encourages others to open up as well.

"A true friend is a keeper of your past, a believer in your future, and a lover of your soul just as it is." – Alex Salem

---

2. Creating Shared Experiences

Shared experiences deepen connections by fostering memories and building a sense of belonging. Examples include:

- **Trying New Activities Together:** Whether it's learning a new skill, taking a trip, or exploring a hobby, shared experiences create lasting bonds.
- **Celebrating Milestones:** Acknowledge and celebrate significant events in each other's lives, such as birthdays, promotions, or personal achievements.
- **Participating in Traditions:** Establish routines or rituals, like weekly coffee dates or annual trips, that reinforce the relationship over time.

"Shared experiences create joyful bonds that last a lifetime." - Alex Salem

3. The Role of Honesty and Boundaries

Authenticity and boundaries are essential for maintaining healthy relationships. Genuine connections thrive on honesty, where individuals feel free to express themselves without fear of rejection or misunderstanding. Simultaneously, setting boundaries ensures mutual respect and prevents emotional burnout:

- **Communicate Needs:** Clearly express what you need from the relationship while being open to hearing the other person's expectations.
- **Respect Boundaries:** Honor personal limits, whether it's about time, space, or emotional capacity.
- **Handle Conflict Gracefully:** Disagreements are inevitable in any relationship. Approach conflicts with an open mind, focusing on resolution rather than blame.

"Honesty builds trust, and boundaries preserve it; together, they create the foundation for healthy relationships." - Alex Salem

Focusing on depth in relationships brings a fulfillment that surface-level interactions cannot. Emotional intimacy, shared experiences, and mutual respect create bonds that endure, enriching our lives and fostering a sense of belonging.

Fostering Relationships in a Digital Era: Navigating Technology's Role

"Technology connects us more than ever, but it's the human touch that makes those connections meaningful." – Unknown

In the digital age, technology has revolutionized how we communicate, offering convenience and immediacy. However, it has also introduced challenges, such as superficial interactions, the illusion of connection, and increased loneliness. Striking a balance between digital tools and authentic relationships is key to fostering meaningful connections in today's world.

1. Recognizing the Limits of Technology

While platforms like social media, video calls, and messaging apps make staying in touch easier, they lack the depth of face-to-face interactions. Digital communication often misses subtle nonverbal cues, such as body language or tone, which are crucial for emotional connection. To combat this:

- **Prioritize In-Person Interactions:** Whenever possible, choose to meet face-to-face rather than relying on digital means.
- **Use Technology as a Supplement:** Let digital tools enhance relationships rather than replace genuine efforts, such as using video calls to stay connected when geographically separated.
- **Be Intentional with Online Interactions:** Move beyond casual likes and comments to meaningful conversations that build rapport.

2. Avoiding the Illusion of Companionship

AI technologies, such as chatbots and virtual assistants, often mimic human interactions, providing a sense of companionship. While these tools can fill temporary gaps, they are not substitutes for genuine relationships. Over-reliance on AI may lead to deeper feelings of isolation. To avoid this trap:

- **Seek Human Connection First:** Engage with friends, family, or community groups to fulfill emotional needs.
- **Use AI as a *Tool,* Not a Crutch:** For example, let AI streamline tasks or provide entertainment but rely on real people for emotional support and meaningful discussions.

"Technology can create the illusion of closeness without the substance of connection." – Sherry Turkle

3. Leveraging Technology to Strengthen Bonds

When used mindfully, technology can enhance relationships by:

- **Maintaining Long-Distance Friendships:** Video calls, messaging apps, and shared online activities (like gaming or streaming movies) help bridge physical gaps.

- **Scheduling Time Together:** Use apps to plan meetups or set reminders for special occasions, ensuring you stay connected despite busy schedules.
- **Sharing Experiences Digitally:** Platforms like social media can be a way to celebrate milestones, share memories, or support each other from afar.

4. Setting Boundaries with Technology

To maintain balance, it's essential to set boundaries for technology usage. For example:

- **Limit Screen Time:** Designate "device-free" zones or times, such as during meals or family gatherings, to foster deeper in-person connections.
- **Practice Digital Detoxes:** Periodically unplug from devices to refocus on relationships and personal well-being.
- **Communicate Boundaries:** If you prefer face-to-face conversations or specific modes of communication, let others know your preferences.

"Human connection cannot be automated."

By understanding technology's strengths and limitations, we can use it to complement rather than replace meaningful relationships. With intentionality, digital tools can become powerful allies in nurturing connections in the modern era.

Strategies for Cultivating Deep and Lasting Connections

"Relationships are built on small, consistent actions over time." – Unknown

Deep, meaningful relationships don't happen overnight; they require effort, consistency, and mutual care. In a world where isolation and superficial connections are increasingly common, it's more important than ever to adopt intentional strategies to nurture genuine bonds. Whether you're seeking to strengthen existing relationships or form new ones, the following approaches can help build trust, empathy, and connection.

---

1. Practice Consistency

"Persistence gets it, and consistency keeps it. Consistency is the key that unlocks the door from average to excellence." – Alex Salem

Consistency is the backbone of strong relationships. Small, regular gestures show that you value the relationship and are willing to invest in it. Examples include:

- Checking in with a friend or loved one weekly, even with a simple text or call.
- Remembering important dates, like birthdays or anniversaries, and celebrating them.

- Following through on commitments, reinforcing reliability and trust.

"In the end, we will remember not the words of our enemies, but the silence of our friends." – Martin Luther King Jr.

2. Focus on Empathy

Empathy is the ability to understand and share another person's feelings, creating a foundation for deeper connections. To practice empathy:

- **Listen Without Interrupting:** When someone shares their thoughts, resist the urge to jump in with advice or solutions. Focus on understanding their perspective.

- **Acknowledge Their Emotions:** Reflect back on what they're feeling by saying, "It sounds like you're really frustrated" or "That must be so exciting."

- **Show Nonverbal Support:** Simple actions like maintaining eye contact, nodding, or offering a comforting touch can convey empathy without words.

"Empathy is being concerned about the human being, not just their output." - Simon Sinek

3. Be Present and Mindful

"Be mindful. Confine yourself to the present; we are only promised today, not tomorrow." — Alex Salem

In today's distraction-filled world, giving someone your undivided attention is a powerful act. When engaging with others:

- Put away devices and eliminate distractions to show you're fully invested in the conversation.
- Practice active listening by summarizing what they say and asking thoughtful questions.
- Be mindful of their nonverbal cues, such as facial expressions or body language, to deepen your understanding of their emotions.

"The greatest gift you can give someone is your presence." – Thich Nhat Hanh

---

4. Resolve Conflicts with Care

No relationship is without challenges, but how you handle conflict can strengthen or weaken a bond. To resolve disputes constructively:

- Address issues early rather than letting resentment build.

- Use "I" statements, such as "I felt hurt when..." to express feelings without blaming the other person.
- Focus on finding solutions rather than assigning fault, emphasizing the value of the relationship over the disagreement.

"Peace is not absence of conflict; it is the ability to handle conflict by peaceful means." - Ronald Reagan

5. Expand Your Circle with Purpose

"There's no limit possible to the expansion of each one of us." Charles M. Schwab

While nurturing existing relationships is vital, forming new connections brings fresh perspectives and opportunities. To build meaningful new relationships:

- Join groups or communities centered on your interests, such as book clubs, sports teams, or volunteering organizations.
- Take the initiative to introduce yourself and start conversations, even if it feels uncomfortable initially.
- Seek out people who align with your values and offer mutual support.

"Strangers are just friends waiting to happen." – Rod McKuen

6. Prioritize Forgiveness and Grace

Holding onto grudges can create emotional barriers that hinder connection. Practicing forgiveness allows relationships to heal and grow. To foster forgiveness:

- Acknowledge your feelings and the impact of the hurt.
- Communicate openly with the person involved, seeking resolution rather than retribution.
- Let go of past grievances to focus on the present and future of the relationship.

"Forgive others, not because they deserve forgiveness, but because you deserve peace." - Mel Robbins

Conclusion: The Joy of Connection

Meaningful relationships are a source of joy, resilience, and fulfillment. They enrich our lives by providing support, love, and shared experiences that no technology or AI can replicate. By practicing consistency, empathy, and mindfulness, we can nurture connections that stand the test of time.

"In the sweetness of friendship let there be laughter, and sharing of pleasures. For in the dew of little things, the heart finds its morning and is refreshed." – Khalil Gibran

Chapter 20: Protecting Your Creativity and Mastering AI

The Intersection of Creativity and AI: Navigating the Tech-Driven Landscape

"The great promise of technology is to expand human potential and foster connections." – Satya Nadella

As artificial intelligence becomes a fundamental part of our daily lives, its influence on creativity and individuality grows. AI-powered tools, from text generators to design platforms, offer unparalleled opportunities for enhancing productivity and generating ideas. Yet, this integration also raises questions about authenticity, originality, and the risk of over-reliance on technology.

At its core, creativity is uniquely human. It involves intuition, emotion, and the ability to draw connections between seemingly unrelated ideas. AI, on the other hand, excels at analyzing patterns, processing vast amounts of data, and offering solutions based on logic. The interplay of these strengths offers a powerful opportunity: AI can amplify creativity by handling routine tasks and providing inspiration, leaving humans free to focus on innovation and personal expression.

However, challenges arise when individuals rely too heavily on AI tools. Over-dependence can dull critical thinking, hinder the development of unique skills, and blur the line between human originality and machine-generated outputs. To preserve your creativity in a tech-driven world, it's essential to approach AI as a collaborator rather than a replacement.

Finding the Balance Balancing AI use with personal creativity begins with questioning its role. Is AI merely a means to an end, or is it becoming something more? When used thoughtfully, it can enhance productivity and support creative endeavors. But when relied upon too heavily, it risks diluting authenticity and reducing human input to a passive, supervisory role.

"The key to artificial intelligence has always been the representation." – Jeff Hawkins

This chapter explores strategies for integrating AI into your creative and professional routines while preserving your unique voice. By mastering this balance, you can harness AI's potential to enhance creativity and productivity without compromising your individuality.

Leveraging AI to Enhance Creativity: Tools and Techniques

"AI can be a great servant, but a terrible master. The key lies in how we choose to use it." – Unknown

Artificial intelligence is a transformative *innovation* that can elevate your creative pursuits by streamlining processes, generating ideas, and offering insights that might otherwise take hours of manual effort. When used thoughtfully, AI complements human ingenuity, enhancing productivity and allowing for deeper exploration of creative possibilities. However, understanding how to incorporate AI effectively is crucial to ensuring that it serves your vision without overshadowing it.

1. Using AI to Streamline Routine Tasks

One of AI's greatest strengths is its ability to handle repetitive and time-consuming tasks. By delegating mundane work to AI, you can focus on high-value, creative activities:

- **Content Creation:** GPT-powered Tools like ChatGPT (OpenAI), Copilot (Microsoft), Gemini (Google), and GPT-4o (OpenAI); writing assistants can draft articles, create social media captions, or brainstorm ideas, freeing up your time for refining and personalizing content.
- **Design Assistance:** AI platforms like Canva or Adobe Sensei offer automated design

suggestions, helping users create visually appealing graphics without starting from scratch.

- **Project Management:** AI-driven tools like Trello or Monday.com streamline workflows by automating task assignments, deadlines, and updates.

This delegation allows you to reserve your mental energy for tasks that require intuition, emotional resonance, and a personal touch.

2. Incorporating AI for Idea Generation

AI excels at analyzing patterns and generating suggestions, making it a valuable partner in brainstorming sessions. Tools like Jasper AI, DALL·E, or ChatGPT can provide prompts, creative variations, or visual concepts based on your input. For example:

- Writers can use AI for story prompts, dialogue ideas, or plot twists.
- Artists can generate inspiration for compositions or experiment with new styles.
- Entrepreneurs can explore product concepts or marketing strategies.

"Inspiration exists, but it has to find you working." – Pablo Picasso

By using AI as a starting point, you gain access to a reservoir of ideas, which you can refine and transform into something truly original.

3. Avoiding Over-Reliance on AI

While AI offers tremendous value, over-reliance can lead to creative stagnation. To prevent this:

- **Set Boundaries:** Use AI for support, but ensure the core ideas and decisions come from you.
- **Practice Manual Creativity:** Regularly engage in activities without AI assistance—writing by hand, sketching, or brainstorming without tools—to maintain your creative instincts.
- **Critically Evaluate AI Outputs:** Avoid taking AI suggestions at face value. Analyze, adapt, and refine them to align with your unique style and goals.

By blending AI's efficiency with your personal creativity, you can maintain control over your work while benefiting from technology's strengths.

Leveraging AI effectively requires a strategic approach, balancing its capabilities with your input. By streamlining tasks, using AI for inspiration, and avoiding over-dependence, you can harness its potential while preserving the authenticity of your creative endeavors.

BEAT THE BOT: Stay in Charge in a World Powered by AI" encourages us to harness the potential of artificial intelligence while maintaining our human agency. It emphasizes that we can coexist with AI, leveraging its capabilities to enhance our lives rather than feeling overwhelmed or replaced by it. By understanding AI and staying informed, we empower ourselves to navigate this AI-driven world effectively.

Preserving Your Unique Voice: Creativity in an AI-Driven World

"Originality thrives in seclusion free of outside influences beating upon us to cripple the creative mind." – Nikola Tesla

As AI becomes more integrated into our daily lives, maintaining your unique voice and creative authenticity is increasingly important. While AI can generate ideas and automate processes, it cannot replicate the personal perspective, emotional depth, and individuality that make human creativity so powerful. Ensuring that your work reflects your identity requires a deliberate effort to balance AI's contributions with your personal touch.

1. Defining Your Creative Identity

Your creative voice reflects your values, experiences, and perspective. To preserve this authenticity:

- **Clarify Your Style:** Take time to identify what makes your work unique. Whether it's a specific tone in writing, a distinct aesthetic in art, or a signature approach in problem-solving, knowing your strengths ensures that AI complements rather than overshadows your originality.
- **Stay True to Your Intentions:** Use AI as a *tool* to express your vision, not to shape it entirely. Before incorporating AI, ask yourself, "Does this align with my core ideas?"
- **Document Your Inspirations:** Maintain a journal or mood board of ideas, inspirations, and influences to ground your work in your own creativity, even when using AI tools.

"Be yourself; everyone else is already taken." – Oscar Wilde

2. Balancing AI Input with Human Insight

AI often excels at generating technically perfect outputs, but it lacks the intuition and emotional nuance that define human creativity. To strike a balance:

- **Layer Personal Insights:** Build on AI-generated ideas with your thoughts, experiences, and feelings. For instance, a writer might use AI to draft a plot outline but infuse it with their personal storytelling style.

- **Focus on Emotional Resonance:** Ensure your work connects with your audience on a human level by adding relatable emotions, cultural context, or personal anecdotes.
- **Review and Refine:** AI-generated outputs are a starting point, not the final product. Take time to edit, adapt, and add layers of depth that reflect your unique perspective.

3. Cultivating Critical Thinking

Creativity thrives when you question assumptions, challenge norms, and explore new ideas. While AI can provide insights and shortcuts, it cannot replicate the critical thinking required for true innovation:

- **Evaluate AI Outputs Thoughtfully:** Assess whether AI-generated suggestions align with your goals, audience, and purpose. Avoid passively accepting results.
- **Experiment Freely:** Use AI to explore unconventional ideas or push boundaries, but don't hesitate to deviate from its recommendations when your instincts guide you elsewhere.
- **Embrace Mistakes:** Unlike AI, humans learn and grow from imperfections. Mistakes are often the catalyst for breakthrough ideas, so don't let the pursuit of AI-enabled perfection stifle your willingness to take risks.

"To live a creative life, we must lose our fear of being wrong." Joseph Chilton Pearce

Preserving your unique voice in an AI-driven world requires conscious effort and intentionality. By defining your creative identity, balancing AI input with human insight, and cultivating critical thinking, you ensure that your work remains a genuine reflection of who you are.

Integrating AI Into Daily Routines Without Sacrificing Individuality

"The goal is to use AI as a catalyst for creativity, not as a substitute for it." – Unknown

Incorporating AI into your daily routines can significantly enhance productivity and creativity, but it's crucial to ensure that these *tools* should serve you rather than define you. Mastering the art of leveraging AI effectively requires setting boundaries, maintaining autonomy, and adopting practices that integrate technology into your workflow without diluting your personal identity.

1. Setting Healthy Boundaries with AI

AI can quickly become a crutch if not used thoughtfully. To avoid over-dependence:

- **Limit AI Use to Specific Tasks:** Identify which tasks benefit most from automation, such as data organization, basic research, or preliminary drafts, and reserve creative decision-making for yourself.

- **Establish Tech-Free Zones:** Designate periods during your day for offline activities. For example, brainstorming sessions, sketching, or journaling without digital tools can foster unfiltered creativity.
- **Audit Your AI Usage Regularly:** Periodically evaluate how and when you use AI. Are these *tools* enhancing your work or unintentionally stifling your creativity?

2. Personalizing AI Outputs

Many AI tools allow customization, making it easier to tailor outputs to align with your unique voice and style. To ensure AI reflects your individuality:

- **Train AI with Personal Input:** Tools like custom GPT models or AI-powered design platforms often let users input preferences, past work, or examples to fine-tune results.
- **Review for Alignment:** Always evaluate AI-generated content against your vision. Make edits that reinforce your personal perspective and tone.
- **Add a Human Touch:** Whether it's adding metaphors in writing, subtle imperfections in artwork, or cultural nuances in marketing materials, small details humanize AI outputs.

3. Blending Technology with Analog Practices

While AI can accelerate many processes, blending technology with traditional methods ensures your creativity remains dynamic and well-rounded. Consider:

- **Alternating Between Digital and Physical Mediums:** If you're an artist, experiment with hand-drawn sketches before digitizing your work. For writers, outline ideas in a notebook before refining them with AI tools.
- **Incorporating Mindful Activities:** Activities like walking, meditating, or even daydreaming stimulate creativity in ways technology cannot replicate. Allow yourself time to disconnect and let your mind wander.

"Your mind is your primary creative tool. AI is just an extension of it." – Anonymous

4. Embracing Collaborative Creativity

Instead of viewing AI as a rival, think of it as a creative partner. Collaboration with AI can:

- **Expand Your Perspectives:** AI often suggests ideas you might not have considered, serving as a catalyst for innovation.
- **Enhance Efficiency:** By taking care of repetitive tasks, AI frees up time for exploring and refining your most inspired ideas.

- **Challenge Your Assumptions:** AI-generated insights can push you to rethink existing concepts, sparking growth and originality.

By integrating AI mindfully into your daily routines, you can harness its power to enhance your creativity and productivity while maintaining a firm grip on your individuality. Thoughtful practices and balanced use ensure that AI remains simply a *tool* that supports your goals rather than defining them.

Mastering AI for Personal Independence and Creative Empowerment

"Artificial intelligence will never replace human ingenuity; it will only magnify it." – Ginni Rometty

As AI continues to reshape the creative and professional landscape, mastering its use is no longer optional; it's essential. The challenge lies in using AI to empower yourself while safeguarding your individuality. By fostering a thoughtful approach to AI, you can maintain control over your creative process, use technology to amplify your capabilities, and ensure that your work reflects your unique voice.

1. Regaining Control Over Your Creative Process

The key to maintaining independence in a technology-driven world is to remain the decision-maker in your creative journey. This requires active involvement at every stage of the process:

- **Define Your Vision:** Before engaging AI, clarify your goals and objectives. Knowing what you want ensures that AI serves your purpose, not the other way around.
- **Stay in the Driver's Seat:** Use AI to handle technical tasks or provide inspiration but retain full control over final decisions and creative direction.
- **Iterate Thoughtfully:** Treat AI outputs as drafts, not finished products. Review, revise, and refine them to align with your vision and standards.

"AI, the new digital species, not only creates but also assists its creator. Our coexistence is essential." – Alex Salem

2. Encouraging Personal Growth Through AI

AI's ability to analyze data, identify trends, and offer suggestions can also be a solution for personal growth. Use AI to:

- **Identify Strengths and Weaknesses:** AI analytics can provide feedback on your work, helping you understand areas for improvement.
- **Explore New Skills:** Platforms that offer AI-powered tutorials, simulations, or guided

learning paths allow you to acquire new knowledge efficiently.

- **Track Progress:** Use tools that monitor your productivity, time management, or creative output to measure growth and set future goals.

By approaching AI as a resource for self-improvement, you turn it into a partner in your personal and creative development.

3. Promoting Originality in a World of Automation

As AI-generated content becomes more prevalent, standing out requires an emphasis on originality and authenticity. To preserve these qualities:

- **Develop a Distinct Style:** Cultivate a creative signature that sets you apart, whether it's a unique writing tone, artistic technique, or approach to problem-solving.
- **Experiment Fearlessly:** Use AI to test unconventional ideas or generate inspiration but take risks beyond what the algorithm suggests. Originality often lies in breaking patterns, not following them.
- **Add Personal Context:** Infuse your work with experiences, emotions, and insights that only you can provide. These elements ensure that your creations remain genuinely human.

"Originality is the art of concealing your sources." – Benjamin Franklin

4. Ensuring Ethical AI Usage

Mastering AI also involves understanding its ethical implications. As a user, you have a role in ensuring that AI serves society responsibly. Consider:

- **Evaluating AI's Impact:** Be aware of how AI influences your field and whether it creates barriers to diversity or access.
- **Using AI Transparently:** Acknowledge when AI contributes to your work to maintain honesty and accountability.
- **Advocating for Fair Practices:** Support initiatives and organizations that promote ethical AI development, ensuring its benefits are widely shared.

5. Achieving a Balance Between Innovation and Tradition

Incorporating AI does not mean abandoning traditional methods. The most innovative creators blend old and new approaches to achieve remarkable results:

- **Respect Legacy Techniques:** Honor traditional practices in your field while exploring how AI can enhance or complement them.
- **Celebrate Human Creativity:** Recognize that AI is just a tool, but creativity originates from the human mind and spirit.

- **Lead by Example:** Show others how to integrate AI thoughtfully, inspiring balanced and ethical use of technology.

"Technology is just a tool. In terms of getting the kids working together and motivating them, the teacher is the most important." – Bill Gates

Conclusion: Creativity in the Age of AI

Mastering AI while protecting your creativity is about balance. By using AI as a support system to enhance your work—not define it—you maintain control over your creative process, preserve your individuality, and push the boundaries of what's possible. The insights and strategies in this chapter empower you to navigate the intersection of technology and creativity, ensuring that your unique voice shines brightly in a tech-driven world.

"The future is not something we enter. The future is something we create." - Leonard Sweet

Conclusion: The Path Forward

Embracing AI and Personal Growth: The Journey Begins

"The future belongs to those who prepare for it today." – Malcolm X

As we stand at the crossroads of technology and humanity, artificial intelligence offers both unprecedented opportunities and complex challenges. This book has explored the many facets of integrating AI into our lives—from enhancing creativity and financial stability to preserving meaningful relationships and fostering personal growth. The central message throughout has been clear: AI is not here to replace human potential; it's a novel digital species—a creator and inventor—crafted and created by us to coexist with us and enhance our future and productivity.

Actively engaging with AI goes beyond comprehending its capabilities; it demands a shift in mindset and the initiation of personal development. Instead of fearing AI's impact, we must approach it as a collaborator in our journey toward self-improvement. When we harness AI with intention, we empower ourselves to navigate the complexities of modern life with resilience, adaptability, and confidence. But this journey is not just about technology; it is also about rediscovering what it means to be human.

Key Takeaways

1. **AI as a Catalyst, not a Replacement:** There is no doubt that AI can certainly enhance efficiency, creativity, and productivity; its true value lies in complementing human intelligence rather than replacing it. Collaborating with AI enables us to maintain control over our distinct contributions. As it evolved, AI has revealed its immense potential to work alongside humans, amplifying our capabilities and creating fresh avenues for collaboration. It is our responsibility to guide this digital companion to become a capable partner—learning, evolving, and contributing to seamless productivity on our terms. By committing to personal development and embracing pillars such as discipline and creativity, we empower ourselves to excel in this dynamic and ever-evolving landscape.
2. **Personal Growth as a Lifelong Commitment:** Having a growth mindset is a revolutionary concept in psychology and education, as it unlocks the transformative belief in our potential for growth. By incorporating discipline and dedication into our daily efforts, we can expand and refine our abilities. Just as muscles grow stronger

through practice and persistence, our skills and intelligence can be developed over time. Therefore, cultivating this mindset within ourselves and encouraging it in others is essential. Setting goals and building strong relationships build a solid foundation for a fulfilling life. Which, in turn, provides us with purpose and stability amidst a fast-paced, tech-driven world.

3. **The Importance of Balance:** Once at the forefront of innovation, the digital landscape has now developed into a complex network of linked platforms, apps, and gadgets. Our communication, work, learning, and perception of the world have all been changed as its consequence. Moreover, the constant barrage of messages and the need to be online at all times is becoming detrimental to our mental health. And for that reason, I believe, it's crucial to take charge and transform into masters rather than being mere users. Maintaining equilibrium is essential to living a peaceful and powerful life, whether it be in managing relationships and work or striking a balance between AI and individual creativity.

"Success is not the key to happiness. Happiness is the key to success. If you love what you are doing, you will be successful." – Albert Schweitzer

The insights shared in this book are not just strategies; they are an invitation to transform how you approach your personal and professional life. As you move forward, remember that your journey is unique. AI may shape the landscape, but your choices define the path.

Reclaiming Your Power: Personal Empowerment in a Tech-Driven World

"In the middle of difficulty lies opportunity." – Albert Einstein

The integration of artificial intelligence into daily life has brought us to a pivotal moment—one where we must choose how to engage with the technologies shaping our world. While AI offers tools that enhance convenience and efficiency, it is ultimately our choices, actions, and mindset that determine whether it serves as an empowering force or an overbearing influence. The path forward is not about passive acceptance but active participation in shaping our future.

Personal Empowerment Through AI

To engage with AI meaningfully, we must take deliberate steps to remain at the center of the decision-making process:

- **Control Over Automation:** AI should work for you, not the other way around. Set boundaries on how and when you use AI to ensure it enhances your life without dictating it.

- **Critical Engagement:** Question the outputs, suggestions, and limitations of AI. By maintaining a curious and critical mindset, you can better leverage AI's strengths while mitigating its weaknesses.
- **Focusing on Growth Areas:** Use AI to identify areas for personal improvement, whether through skill-building, creative pursuits, or managing resources more effectively. These applications of AI empower you to become the best version of yourself.

"The question of whether machines can think is about as relevant as the question of whether submarines can swim." – Edsger Dijkstra

The Role of Human Connection

Amid rapid technological change, the need for authentic human connections has never been more critical. This book has underscored the value of relationships—both personal and professional—as a cornerstone of well-being and growth. To balance the isolating tendencies of digital tools:

- Prioritize face-to-face interactions whenever possible.
- Engage in activities that nurture emotional intimacy, such as shared experiences and open conversations.

- Use AI sparingly in social contexts, ensuring that technology does not replace genuine connections.

Meaningful relationships are the antidote to the alienation that can accompany a tech-driven society. By cultivating these bonds, you create a support network that grounds and sustains you through challenges.

Building Resilience in Uncertainty Life's unpredictability is compounded by the rapid pace of technological and societal change. Resilience is no longer optional; it's a necessity for navigating these complexities. Whether it's adapting to workplace disruptions caused by AI, managing financial shifts, or facing personal challenges, the strategies discussed in this book equip you to remain strong and adaptable. Building resilience involves:

- A commitment to lifelong learning.
- A willingness to embrace change as an opportunity for growth.
- A mindset that prioritizes self-care, emotional balance, and long-term planning.

"You may not control all the events that happen to you, but you can decide not to be reduced by them." – Maya Angelou

As you reflect on the strategies shared in this book, remember that personal empowerment is an ongoing process. The choices you make today will shape the opportunities of tomorrow, allowing you to thrive in a world increasingly influenced by AI.

Stepping into the Future: Your Role in Shaping What Comes Next

"Education is the passport to the future, for tomorrow belongs to those who prepare for it today." - Malcolm X:

The journey I've embarked upon in this book pulses with urgency—a call to action. As AI reshapes our world, your role as an active participant becomes paramount. The choices you make, the strategies you implement, and the mindset you adopt will determine whether you thrive in this tech-driven era or become overwhelmed by it.

Embrace the Pillars of Personal Growth:

Till this point, we have discussed the importance of personal growth, but how we can achieve that has been disucced below.

Discipline: Like the steady beat of a metronome, discipline provides rhythm to progress. It's an unbreakable commitment to daily habits and the relentless pursuit of excellence. Whether it's dedicating time to learn a new skill, maintaining a fitness routine, or consistently honing your craft, discipline fuels forward momentum.

Creativity: Think of your mind as an empty canvas that is just awaiting the strokes that will turn it into a great work of art. Your capacity to think creatively and connect seemingly unrelated dots becomes a strength in a world where algorithms generate predictable patterns. Your creative potential can be unlocked and enhanced by a variety of characteristics, including an unyielding curiosity, an adventurous spirit, and the courage to dream large.

Resilience: Life's storms may buffet you, but resilience is an element that anchors you. It's the oak tree that bends but doesn't break. When setbacks come—whether in your career, relationships, or personal endeavors—resilience whispers, "This too shall pass." It's an unwritten chapter of your story, where setbacks become stepping stones, propelling you to strive harder and move forward.

Leverage the new digital species AI as Your Co-Creator:

AI isn't an adversary; it's a co-creator. Imagine it as a symphony conductor, weaving harmonies alongside your melodies. Let AI handle repetitive tasks, analyze data, and predict trends. But remember, the baton is in your hands. You are the one who has the power to decide the tempo, the crescendos, and the pauses. AI amplifies your reach, but your intuition guides the entire composition.

Forge Ahead with Purpose:

The future is the current that carries you forward, not some far-off shore. Accept uncertainty as a friend. Don't be afraid to innovate. Work together across disciplines. Look for mentors who have experienced new situations. Additionally, keep in mind that you are not alone when the waves of doubt smash. Millions of artificial and human minds are shaping the tides.

Enter the future: a scientist with a soul and an artist with algorithms. Code your syntax; the canvas is waiting for your brushstroke. That is your legacy, your codebase, and your symphony. Let's design a world in which advancement moves in unison with purpose and technology benefits humanity.

Living with Intention in an AI-Driven World

Adopting a mindset of intentionality is key to navigating the future with confidence. Living with intention means:

- **Defining Your Values:** Know what matters most to you—whether it's creativity, relationships, independence, or financial stability—and let these values guide your decisions.
- **Using AI Purposefully:** Treat AI as a means to amplify your strengths, not as a shortcut that diminishes your role in your success.

Always ask, "How can this *tool* help me achieve my goals?"

- **Balancing Innovation with Authenticity:** Embrace technological advancements while staying true to your unique voice and personal contributions. Let AI enhance your work, but ensure it reflects your individuality.

"Don't let the noise of others' opinions drown out your own inner voice." – Steve Jobs

The Ripple Effect of Personal Growth

Personal development is not just about individual success—it has a ripple effect that extends to your relationships, community, and society at large. As you apply the strategies in this book to improve your life, you also contribute to a better world:

- **Inspiring Others:** Your commitment to growth, resilience, and creativity serves as a model for others to follow, encouraging them to take charge of their own journeys.

- **Strengthening Communities:** By fostering authentic connections and supporting those around you, you build stronger, more compassionate communities.

- **Shaping the Future:** Your choices influence how AI is adopted and integrated, ensuring it serves humanity's best interests.

"Be the change you wish to see in the world." – Mahatma Gandhi

Your Path Forward

As you step into the future, remember that the journey is ongoing. The insights shared in this book are starting points for continuous exploration and growth. Use them to:

- Adapt to challenges with resilience and creativity.
- Pursue your passions with clarity and purpose.
- Cultivate relationships that enrich your life.
- Leverage AI as an entity for empowerment, as a partner, not dependency.

A Final Thought

"The future grows from today's seeds." – Alex Salem

The path forward is yours to define. While the world may change around you, the principles of growth, authenticity, and connection remain timeless. By embracing these values, you ensure that technology serves you, not the other way around. Your future is not a destination; it's a series of intentional choices that shape your life and the world around you.

"Be built, not born. If you want to inherit the future, you must be disciplined in the present– Alex Salem

As this book concludes, let me share some of the good habits that have brought me to where I am today. I hope and believe my readers can draw inspiration from them and work toward their own growth and success.

"Wake up, drill up, clean up, dress up, show up, and never give up." – Alex Salem

Wake up early to start your day with intention.

Exercise to boost your physical and mental health.

Meditate for inner peace and clarity.

Read to expand your knowledge and imagination.

Take a cold shower to improve circulation and better mood.

Plan your day to stay organized and productive.

Hydrate by drinking plenty of water.

Eat a healthy breakfast to fuel your body.

Practice gratitude to foster a positive mindset.

Set daily goals to stay focused and motivated.

Take breaks to recharge and maintain productivity.

Learn something new to keep your mind sharp.

Connect with loved ones to strengthen relationships.

Practice mindfulness to stay present and engaged.

Declutter your space to create a more organized environment.

Journal to reflect on your thoughts and experiences.

Limit screen time to reduce eye strain and improve sleep.

Get outside to enjoy fresh air and nature.

Help others to foster a sense of community and purpose.

Review your day to assess your progress and make adjustments.

Get enough sleep to ensure your body and mind are well-rested.

Practice deep breathing to reduce stress and anxiety.

Listen to music to uplift your mood.

Smile to improve your mood and the mood of others.

Practice positive affirmation to boost your confidence.

Stretch to improve flexibility and reduce tension.

Cook a healthy meal to nourish your body.

Limit caffeine intake to avoid jitters and improve sleep.

Practice good posture to prevent back pain and improve confidence.

Engage in a hobby to relax and enjoy yourself.

Network to build professional relationships.

Practice self-care to maintain your well-being.

Visualize your goals to stay motivated and focused.

Track your habits to monitor your progress.

Practice forgiveness to let go of grudges and find peace.

Spend time with pets to reduce stress and increase happiness.

Practice empathy to improve your relationships.

Unplugging technology to reconnect with the real world.

Read and create inspirational quotes to stay motivated.

Practice time management to make the most of your day.

Engage in creative activities to stimulate your mind.

Practice financial management to stay on top of your finances.

Volunteer to give back to your community.

Practice patience to improve your interactions with others.

Stay curious to keep learning and growing.

Practice humility to stay grounded and open-minded.

Celebrate small wins to stay motivated.

Practice resilience to bounce back from setbacks.

Stay informed to be aware of the world around you.

Practice kindness to make the world a better place.

Reflect on your values to stay true to yourself.

Incorporating these habits can help you lead a more balanced, productive, and fulfilling life. See which ones resonate the most.

With that, here are some quotes that have always inspired me and that I think can become your source of motivation as well.

1. Achievement – "Success is the sum of small efforts repeated daily."
2. Ambition – "Ambition is the path to success; persistence is the vehicle."
3. Aspire – "Aspire not to have more, but to be more."
4. Balance – "Balance is not something you find, but something you create."
5. Belief – "Belief in yourself is the first step toward success."
6. Boldness – "Fortune favors the bold."
7. Bravery – "Bravery is not the absence of fear, but the strength to move forward despite it."
8. Challenge – "The greater the challenge, the greater the reward."

9. Commitment – "Commitment is what transforms a promise into reality."

10. Confidence – "Confidence is silent. Insecurity is loud."

11. Courage – "Courage is being scared to death but saddling up anyway."

12. Creativity – "Creativity is intelligence having fun."

13. Dedication – "Dedication is the key that unlocks the door to success."

14. Determination – "Determination today leads to success tomorrow."

15. Discipline – "Discipline is the bridge between goals and accomplishment."

16. Dream – "Dream big. Start small. Act now."

17. Drive – "Drive is the fuel that propels you toward your goals."

18. Empower – "Empower others, and you empower yourself."

19. Endurance – "Endurance is the ability to withstand adversity and still keep going."

20. Energy – "Energy flows where attention goes."

21. Enthusiasm – "Enthusiasm is the spark that lights the fire of achievement."

22. Excellence – "Excellence is not a skill, it's an attitude."

23. Faith – "Faith is taking the first step even when you don't see the whole staircase."
24. Focus – "Focus on the possibilities, not the problems."
25. Fortitude – "Fortitude is the courage to endure difficult times with grace."
26. Fulfillment – "Fulfillment comes not from having, but from becoming."
27. Goal – "A goal without a plan is just a wish."
28. Growth – "Growth is never by mere chance; it is the result of forces working together."
29. Hope – "Hope is the thing with feathers that perches in the soul."
30. Imagination – "Imagination is the beginning of creation."
31. Inspiration – "Inspiration is everywhere; you just need to be open to it."
32. Integrity – "Integrity is doing the right thing when no one is watching."
33. Joy – "Joy is not in things; it is in us."
34. Journey – "The journey of a thousand miles begins with a single step."
35. Kindness – "Kindness is a language the deaf can hear and the blind can see."

36. Leadership – "Leadership is not about being in charge, but about taking care of those in your charge."
37. Learning – "Learning is a treasure that will follow its owner everywhere."
38. Love – "Love is the only force capable of transforming an enemy into a friend."
39. Mastery – "Mastery is not the end goal; it is a never-ending journey."
40. Motivation – "Motivation is what gets you started. Habit is what keeps you going."
41. Optimism – "Optimism is the faith that leads to achievement."
42. Passion – "Passion is the fuel in the fire of action."
43. Patience – "Patience is not the ability to wait, but how you act while you're waiting."
44. Perseverance – "Perseverance is the hard work you do after you get tired of doing the hard work you already did."
45. Positivity – "A positive attitude causes a chain reaction of positive thoughts, events, and outcomes."
46. Potential – "Your potential is endless; the only limit is the one you set."
47. Progress – "Progress is the sum of small efforts, repeated day in and day out."

48. Purpose – "Purpose is the reason you journey; passion is the fire that lights the way."
49. Resilience – "Resilience is the ability to bounce back, no matter how hard you fall."
50. Success – "Success is stumbling from failure to failure with no loss of enthusiasm."

I hope you find these words motivating! Which one resonates with you the most? Embrace it, live by it, and let it transform your life.

: Alex Salem "Alexism @alexism365

: Alex Salem "Alexism"

: Alex Salem "Alexism" @alexsalemalexsim

: Alex Salem "Alexism" @alexsalem.alexsim

: Alex Salem "Alexism" @AlexSalemNYC

: Alex Salem "Alexism" @Alexism

Made in the USA
Monee, IL
05 November 2026

67911522R00164